T0305620

TELECOMMUNICATIONS SYSTEM RELIABILITY ENGINEERING, THEORY, AND PRACTICE

TELECOMMUNICATIONS SYSTEM RELIABILITY ENGINEERING, THEORY, AND PRACTICE

Mark L. Ayers

IEEE Press
Series On
Network
Management

Thomas Plevyak and Veli Sahin, *Series Editors*

IEEE PRESS

A JOHN WILEY & SONS, INC., PUBLICATION

Cover Image: Bill Donnelley/WT Design

Copyright © 2012 by the Institute of Electrical and Electronics Engineers, Inc.

Published by John Wiley & Sons, Inc., Hoboken, New Jersey. All rights reserved.
Published simultaneously in Canada.

For general information on our other products and services or for technical support, please contact our
Customer Care Department within the United States at (800) 762-2974, outside the United States
at (317) 572-3993 or fax (317) 572-4002.

Wiley also publishes its books in a variety of electronic formats. Some content that appears in print
may not be available in electronic formats. For more information about Wiley products, visit our web site
at www.wiley.com.

Library of Congress Cataloging-in-Publication Data:

Ayers, Mark L.
  Telecommunications system reliability engineering, theory, and practice / Mark L. Ayers.
    p. cm.
  ISBN 978-1-118-13051-3 (hardback)
  1. Telecommunication systems.   I. Title.
  TK5101.A89 2012
  621.382–dc23   2012013009

Printed in the United States of America

10 9 8 7 6 5 4 3 2 1

CONTENTS

LIST OF ILLUSTRATIONS

PREFACE

The topic of reliability is somewhat obscure within the field of electrical (and ultimately communications) engineering. Most engineers are familiar with the concept of reliability as it relates to their automobile, electronic device, or home, but performing a rigorous mathematical analysis is not always a comfortable or familiar task. The quantitative treatment of reliability has a long-standing tradition within the field of telecommunications dating back to the early days of Bell Laboratories.

Modern society has developed an insatiable dependence on communication technology that demands a complete understanding and analysis of system reliability. Although the technical innovations developed in modern communications are astonishing engineering marvels, the reliability analysis of these systems can sometimes be treated as a cursory afterthought. Even in cases where analysis of system reliability and availability performance is treated with the highest concern, the sophistication of analysis techniques is frequently lagging behind the technical development itself.

The content in this book is a compilation of years of research and analysis of many different telecommunications systems. During the compilation of this research, two primary points became evident to me. First, most communications engineers understand the need for reliability and availability analysis but lack the technical skill and knowledge to execute these analyses confidently. Second, modern communications network topologies demand an approach to analysis that goes beyond the traditional reliability block diagram and exponential distribution assumptions. Modern computing platforms enable engineers to exploit analysis techniques not possible in the days when the Bell Laboratories' techniques were developed and presented. This book presents techniques that utilize computer simulation and random variable models not feasible 20 years ago. I hope that readers of this book find within it a useful resource that I found absent in the academic literatures during my research and analysis of communications system reliability. Although compilation of the data in this book took me years, it is my desire to convey this information to the reader in a matter of hours, enabling engineers to analyze complex problems using basic tools and theories.

I would like to thank Tom Plevyak and Veli Sahin for their editing and review of this book. Their help in producing this book has been instrumental to its completion and quality.

I would also like to thank Gene Strid for his contributions to my career and to the development of this book. His mentoring spirit and attention to detail have had a significant influence on my personal development as a professional engineer. Gene's technical review of this book alone is impressive in its detail and breadth. Thank you, Gene, for everything you have done to help me remain inspired to grow and learn as an engineer and a leader.

ABOUT THE AUTHOR

Mark Ayers is the Manager of RF Engineering at GCI Communications Corporation headquartered in Anchorage, Alaska. Mark has a broad range of telecommunications experience including work in fiber optics, microwave radio, and satellite network designs. Mark holds a B.S. degree in Mathematics from the University of Alaska Anchorage, and an M.S. degree in Electrical Engineering from the University of Alaska Fairbanks, Fairbanks, Alaska. He is a registered Professional Electrical Engineer in the State of Alaska and a Senior Member of the IEEE. Mark teaches a variety of courses as an Adjunct Faculty Member in the Engineering Department at the University of Alaska Anchorage. His primary interests are systems design, modeling, and optimization.

ACRONYM LIST

AC   Alternating current
ACM   Adaptive coding and modulation
AGM   Absorbed glass mat
AP   Access point
AuC   Authentication center
BLSR   Bidirectional line switched ring
BSC   Base station controller
BTS   Base transceiver station
BTU   British thermal unit
BUC   Block upconverter
CDF   Cumulative distribution function
CDMA   Code division multiple access
COTS   Commercial off the shelf
CPE   Customer premise equipment
CRAC   Computer room air conditioner
DC   Direct current
EDFA   Erbium-doped fiber amplifier
EIR   Equipment identity register
EIRP   Equivalent isotropic radiated power
FCC   Federal Communications Commission
FITs   Failures in time
FMEA   Failure mode and effects analysis
FPGA   Field-programmable gate array
FSS   Fixed satellite system
GSM   Global system for mobile communications
HLR   Home location register
HVAC   Heating, ventilation, and air conditioning
IDU   Indoor unit
IEEE   Institute for Electrical and Electronics Engineers
ISM   Industrial, scientific, and medical
ITU   International Telecommunications Union
LHS   Lefthand side
LNA   Low-noise amplifier
LNB   Low-noise block
LTE   Line-terminating equipment

LTE	Long-term evolution
MDT	Mean downtime
MODEM	Modulator/demodulator
MSC	Mobile switching center
MTBF	Mean time between failures
MTTF	Mean time to failure
MTTR	Mean time to repair
NASA	National Air and Space Administration
NSS	Network switching subsystem
OC-n	Optical carrier, level n
ODU	Outdoor unit
PDF	Probability density function
PFE	Power feed equipment
PM	Preventative maintenance
RBD	Reliability block diagram
RF	Radio frequency
RHS	Righthand side
RMA	Return material authorization
RSL	Received signal level
SDH	Synchronous digital hierarchy
SES	Severely error second
SLA	Service-level agreement
SLTE	Submarine line-terminating equipment
SMS	Short message service
SONET	Synchronous optical network
SP	Service provider
SRGM	Software reliability growth model
SSPA	Solid state power amplifier
TDM	Time domain multiplexing
TRX	Transceiver
TTF	Time to failure
TTR	Time to repair
UMTS	Universal mobile telecommunications system
UPS	Uninterruptable power supply
UPSR	Unidirectional path switched ring
VLR	Visitor location register
VLSI	Very large-scale integration
VRLA	Valve-regulated lead acid
VSAT	Very-small-aperture terminal
WiFi	Wireless fidelity
XPIC	Cross-polarization interference cancellation

INTRODUCTION

The concept of reliability is pervasive. It affects our attitudes and impacts our decisions on a daily basis. Its importance would imply that everyone has a clear understanding of reliability from a technical perspective. Unfortunately, the general public typically equates emotion and perception with reliability. In many cases, even technically minded people do not have a clear, quantitative understanding of reliability as a measure of performance.

Reliability engineering is a relatively new field. Although the term reliability has a long history, it was not until the twentieth century that reliability began to take on a quantitative meaning. In the early twentieth century, the concept of reliability engineering began to take form as the industrial revolution brought about mechanical and electronic systems such as the automobile and the telegraph. Large-scale production resulted in an increased awareness of item failure and performance and its impact on business. During the 1930s, Wallodie Weibull began documenting his work on the measurement and definition of material fatigue behavior. The result of his work is the Weibull distribution, one of the most widely used statistical distributions in reliability engineering. The Second World War brought about the formalization of reliability engineering as a field of study. The advent of radar and other electronic

Telecommunications System Reliability Engineering, Theory, and Practice, Mark L. Ayers.
© 2012 by the Institute of Electrical and Electronics Engineers, Inc. Published 2012 by John Wiley & Sons, Inc.

warfare systems identified further the need to begin quantifying reliability and its impacts on mission success. During the Second World War, vacuum tubes were extensively used in many electronic systems. The low reliability of early vacuum tubes led to both poor system performance and high maintenance costs. The IEEE Reliability Society was formed in 1948 as a result of the increasing focus on reliability in electronic systems.

Following the Second World War, reliability engineering began to find applications in both military and commercial environments. System reliability was studied from a life-cycle standpoint including component design, quality control, and failure analysis. Space exploration in the 1960s continued the need for a life-cycle approach to reliability engineering. The establishment of NASA and an interest in nuclear power generation became driving forces for the development of highly reliable components and systems. Launching commercial communications satellites by INTELSAT and landing on moon by the United States proved the importance of reliability engineering as part of the system engineering process at the end of the 1960s. Semiconductor development, military applications, communications systems, biomedical research, and software-based systems in the 1980s led to new work in both system design and reliability analysis. Improved component design and quality control led to significant improvements in reliability performance. Consumer awareness and commercial focus in the 1990s and 2000s led to the current state of reliability engineering in today's society. Most consumers are unconsciously aware of reliability as a measure of an item's performance and overall value. Engineers and technical resources are aware of an item's reliability in a more quantitative sense but many times this understanding is neither complete nor found in solid reliability engineering principles.

The presentation of reliability data, whether qualitative or quantitative, must be based in solid theory. In many cases, reliability data is used to make business and technical decisions with far-reaching implications. Predictive analysis is typically the first step in the reliability engineering process. Target performance measures are used to guide the design process and ensure that system design is compliant with system performance targets. Modern predictive reliability analysis utilizes statistical modeling of component failures. These statistical models are used to predict a number of expected system performance measures. Changing the system topology or design and reanalyzing system performance allows engineering to do cost/performance trade-off analyses. The analyses can then be used to make business and technical decisions about the best design that meets target requirements.

Once a design has been selected and constructed, it is important to collect empirical data. This data allows the engineer or the operator to measure system performance and compare that performance with expected or predicted data. Empirical data collection is particularly important in large production environments where statistical behavior can be observed. These observations can be tabulated and compared with the predicted or assumed behavior, refining the system model and improving future predictions and decisions. In some cases, empirical data can be directly used to analyze the predicted performance of a new system. One must be careful when using empirical data for predictive analysis because it is rare to find an existing system that exactly matches a new design.

One of the most significant benefits of empirical analysis and data collection is failure mode and effects analysis (FMEA). This analysis approach allows the engineer to identify systemic problems and design flaws by observing the failure of components or systems, using this data to improve future performance. Operational models and processes can be adjusted based on failure data and root cause analysis.

Telecommunications systems have a long history of reliability-based design. These design criteria are typically specified in terms of availability, rather than reliability. Availability is another measure of statistical system performance and is indicative of a system's "uptime" or available time for service delivery. In many cases, service contracts or service-level agreements (SLAs) are specified in terms of availability. Service providers (SPs) will sign a contract to provide a service that has specific target probability of being available or a target maximum downtime over a specific time interval. Both of these measures are metrics of availability. Without predictive and/or empirical data to ensure compliance with these targets, the SP and the customer will take risk in signing the contract. This risk is sometimes realized risk (the party is aware of the risk, quantified or not) or unrealized risk (the party is taking risk and is not aware that they are in jeopardy). Decisions made while assuming unrealized risk can jeopardize business. Reliability engineering of systems in telecommunications serves to reduce overall risk in both realized and unrealized cases.

Conducting business in the field of telecommunications always involves making decisions with financial implications. Telecommunications contracts are often written around SLAs in which a performance target is specified. SPs must ensure that their service can achieve the required performance while customers must maintain realistic expectations from the service requested. Without access to a quantitative reliability analysis, these financial decisions are based on assumptions at best and perception at worst. Rigorous reliability engineering and analysis of telecommunications systems allows managers and technical resources to design systems that achieve the required targets with minimum cost and maximum performance.

Analysis of telecommunications systems requires specialized application of reliability engineering theory and principles. Performance expectations within the field of telecommunications can range from high to extreme. Rarely do consumers of telecommunications expect less than highly available systems. This is true even of consumer services such as cable television, consumer Internet, and local telephone service. Commercial service expectations are typically higher than those in a consumer environment because the impact on the business may be significant and costly if their telecommunications services are critical to their operations, delivery of service, and ability to generate revenues. Performing detailed analyses of systems, both consumer and commercial, allow risks to be managed and costs to be controlled. These analyses allow the designer to produce a system that is carefully crafted to just meet the requirements of the customer rather than greatly exceed them or completely miss the target. In the case of operational systems, knowledge of the achievable system performance and its maintainability allows the operator to understand whether their achieved performance is within specification and to optimize maintenance and repair efforts.

This book is written with the goal of providing the reader with the knowledge and skills necessary to perform telecommunications system reliability analysis and to examine system designs with a critical eye. Telecommunications service providers frequently provide service to customers who know what they would like to purchase, whether it is wireless or terrestrial, packet or TDM. It is far less frequent that the customer understands how to specify system availability or reliability. Knowledge of the theory and practice of reliability engineering allows service providers and engineers to educate their customers regarding this important metric of network performance. Even if the reader does not perform firsthand reliability analysis, the knowledge gained by studying both the theory and the practice of reliability engineering allows the individual to make more informed, better decisions about design and operation of telecommunications systems or the purchase of telecommunications services. The truly pervasive nature of reliability, as a metric in telecommunications systems, requires engineers, managers, and executives to have extensive knowledge of system topologies, costs, and performance. In many cases, these system details are obtained through experience and practice. The author of this book would argue that experience without academic study, particularly in the field of reliability engineering, results in decisions that at times invoke unrealized, serious business risk.

The reader is expected to have a basic working knowledge of engineering mathematics. A college-level course in probability and statistics is of particular value to the reader. This book relies extensively on the application and use of statistical distributions and probability models. Experience with telecommunications system design and network topologies is valuable in understanding the trade-offs involved with different reliability analyses. Lastly, if the reader has interest in developing his or her own reliability models, knowledge of MATLAB and computer programming methods is of value. All of the topics presented in this book are intended to provide sufficient depth to enable the reader to either work with them directly or conduct minimal further research in order to obtain a complete understanding of a topic.

The previous paragraph should allow readers to identify themselves as a member of a specific group. These groups can generally be classified as one of the following: engineers, managers, or executives. Engineers can use this book as a complete technical resource to be used in building and analyzing system models. The engineer reader that uses this book will have the ability to develop complex, detailed statistical models of telecommunications systems that produce a variety of system metrics that can be used for business, design, and other technical decisions. Managers reading this book will derive value from the knowledge obtained about proper reliable system design, contract implications, and operational impacts. Executive readers will find value in the high-level knowledge obtained about design, best practices, and proper expectations for system performance.

This book is logically organized to provide two distinct sets of information, theory and applications. Chapter 1 introduces and develops the concepts and accepted theories required for system reliability analysis. This includes discussions of probability and statistics, system reliability theory, and system modeling. The remaining chapters of this book are organized by technology subject matter. Chapter 2 discusses fiber-optic networks. Both terrestrial and submarine networks are discussed with the subtleties of each presented in detail. Chapter 3 presents reliability analysis approaches for terrestrial microwave systems. The discussion includes short-haul point-to-point, long-haul point-to-point, cellular wireless, and WiFi networks. Satellite communications networks are discussed in Chapter 4. Both teleport and VSAT network topologies are discussed along with propagation availability calculation techniques. Chapter 5 addresses reliability concerns for mobile wireless (cellular) systems. In Chapter 6 the often underanalyzed topics of power systems and heating, ventilation, and air conditioning systems, related to communications networks, are analyzed. The final chapter (Chapter 7) introduces software and firmware as they relate to telecommunications system reliability. Each section presents the analysis in terms of two discrete parts. These parts are the communications equipment and the communications channel. The goal of this book is to provide the reader with sufficient knowledge to abstract and apply the concepts presented to their own problem statement.

The ability to blend academic theory and practical application is a rare commodity in the field of engineering. Few practicing engineers have the ability to apply abstract theory to real problems while even fewer academics have the practical experience to understand the engineering of "real" systems. Telecommunications reliability engineering necessitates the blend of abstract statistical theory and practical engineering experience. Fortunately, in the case of reliability engineering, this blend is easily understood when the information required is presented in a logical, organized format. The use of predictive and/or numerical models in the design of telecommunications systems brings great value to system designs. Acceptance of these models requires the engineer, manager, and executive to have enough confidence in the model's results so that significant decisions can be made based on the results of that model. The ability to place that level of confidence in a model can only come from a fusion of reliability engineering academics and experience.

1

RELIABILITY THEORY

A solid foundation in theoretical knowledge surrounding system reliability is fundamental to the analysis of telecommunications systems. All modern system reliability analysis relies heavily on the application of probability and statistics mathematics. This chapter presents a discussion of the theories, mathematics, and concepts required to analyze telecommunications systems. It begins by presenting the system metrics that are most important to telecommunications engineers, managers, and executives. These metrics are the typical desired output of an analysis, design, or concept. They form the basis of contract language, system specifications, and network design. Without a target metric for design or evaluation, a system can be constructed that fails to meet the end customer's expectations. System metrics are calculated by making assumptions or assignments of statistical distributions. These statistical distributions form the basis for an analysis and are crucial to the accuracy of the system model. A fundamental understanding of the statistical models used in reliability is important. The statistical distributions commonly used in telecommunications reliability analysis are presented from a quantitative mathematical perspective. Review of the basic concepts of probability and statistics that are relevant to reliability analysis are also presented.

Having developed a clear, concise understanding of the required probability and statistics theory, this chapter focuses on techniques of reliability analysis. Assumptions

Telecommunications System Reliability Engineering, Theory, and Practice, Mark L. Ayers.
© 2012 by the Institute of Electrical and Electronics Engineers, Inc. Published 2012 by John Wiley & Sons, Inc.

adopted for failure and repair of individual components or systems are incorporated into larger systems made up of many components or systems. Several techniques exist for performing system analysis, each with its own drawbacks and advantages. These analysis techniques include reliability block diagrams (RBDs), Markov analysis, and numerical Monte Carlo simulation. The advantages and disadvantages of each of the presented approaches are discussed along with the technical methodology for conducting each type of analysis.

System sparing considerations are presented in the final section of this chapter. Component sparing levels for large systems is a common consideration in telecommunications systems. Methods for calculating sparing levels based on the RMA repair period, failure rate, and redundancy level are presented in this section.

Chapter 1 makes considerable reference to the well-established and foundational work published in "System Reliability Theory: Models, Statistical Methods and Applications" by M. Rausand and A. Høyland. References to this text are made in Chapter 1 using a superscript[1] indicator.

1.1   SYSTEM METRICS

System metrics are arguably the most important topic presented in this book. The definitions and concepts of reliability, availability, maintainability, and failure rate are fundamental to both defining and analyzing telecommunications systems. During the analysis phase of a system design, metrics such as availability and failure rate may be calculated as predictive values. These calculated values can be used to develop contracts and guide customer expectations in contract negotiations.

This section discusses the metrics of importance in telecommunications from both a detailed technical perspective and a practical operational perspective. The predictive and empirical calculation of each metric is presented along with caveats associated with each approach.

1.1.1   Reliability

MIL-STD-721C (MILSTD,1981) defines reliability with two different complementary definitions.

1. The duration or probability of failure-free performance under stated conditions.
2. The probability that an item can perform its intended function for a specified interval under stated conditions. (For nonredundant items, this is equivalent to definition 1. For redundant items this is equivalent to the definition of mission reliability.)

Both MIL-STD-721C definitions of reliability focus on the same performance measure. The probability of failure-free performance or mission success refers to the likelihood that the system being examined works for a stated period of time. In order to quantify and thus calculate reliability as a system metric, the terms "stated period" and "stated conditions" must be clearly defined for any system or mission.

The stated period defines the duration over which the system analysis is valid. Without definition of the stated period, the term reliability has no meaning. Reliability is a time-dependent function. Defining reliability as a statistical probability becomes a problem of distribution selection and metric calculation.

The stated conditions define the operating parameters under which the reliability function is valid. These conditions are crucial to both defining and limiting the scope under which a reliability analysis or function is valid. Both designers and consumers of telecommunications systems must pay particular attention to the "stated conditions" in order to ensure that the decisions and judgments derived are correct and appropriate.

Reliability taken from a qualitative perspective often invokes personal experience and perceptions. Qualitative analysis of reliability should be done as a broad-brush or high-level analysis based in a quantitative technical understanding of the term. In many cases, qualitative reliability is defined as a sense or "gut feeling" of how well a system can or will perform. The true definition of reliability as defined in MIL-STD-721C is both statistical and technical and thus any discussion of reliability must be based in those terms.

Quantitative reliability analysis requires a technical understanding of mathematics, statistics, and engineering analysis. The following discussion presents the mathematical derivation of reliability and the conditions under which its application are valid with specific discussions of telecommunications systems applications.

Telecommunications systems reliability analysis has limited application as a useful performance metric. Telecommunications applications for which reliability is a useful metric include nonrepairable systems (such as satellites) or semirepairable systems (such as submarine cables). The reliability metric forms the foundation upon which availability and maintainability are built and thus must be fully understood.

1.1.1.1 The Reliability Function.
The reliability function is a mathematical expression analytically relating the probability of success to time. In order to completely describe the reliability function, the concepts of the state variable and time to failure (TTF) must be presented.

The operational state of any item at a time t can be defined in terms of a state variable $X(t)$. The state variable $X(t)$ describes the operational state of a system, item, or mission at any time t. For the purposes of the analysis presented in this section, the state variable $X(t)$ will take on one of two values.[1]

$$X(t) = \begin{cases} 1 & \text{if the item state is functional or successful} \\ 0 & \text{if the item state is failed or unsuccessful} \end{cases} \tag{1.1}$$

The state variable is the fundamental unit of reliability analysis. All of the future analyses will be based on one of two system states at any given time, functional or failed ($X(t) = 1$ or $X(t) = 0$). Although this discussion is limited to the "functional" and "failed" states, the analysis can be expanded to allow $X(t)$ to assume any number of different states. It is not common for telecommunications systems to be analyzed for partial failure conditions, and thus these analyses are not presented in this treatment.

We can describe the operational functionality of an item in terms of how its operational state at time t translates to a TTF. The discrete TTF is a measure of the amount of time elapsed before an item, system, or mission fails. It should be clear that the discrete, single-valued TTF can be easily extended to a statistical model. In telecommunications reliability analysis, the TTF is almost always a function of elapsed time. The TTF can be either a discrete or continuous valued function.

Let the time to failure be given by a random variable T. We can thus write that probability that the time to failure T is greater than $t = 0$ and less than a time t (this is also known as the CDF $F(t)$ on the interval $[0,t)$) as[1]

$$F(t) = Pr(T \leq t) \quad \text{for } [0, t) \tag{1.2}$$

Recall from probability and statistics that the CDF can be derived from the probability density function (PDF) by evaluating the relationship

$$F(t) = \int_0^t f(u)\, du \quad \text{for all } t \geq 0 \tag{1.3}$$

where $f(u)$ is the PDF of the time to failure. Conceptually, the PDF represents a histogram function of time for which $f(t)$ represents the relative frequency of occurrence of TTF events.

The reliability of an item is the probability that an item does not fail for an interval $(0, t]$. For this reason, the reliability function $R(t)$ is also referred to as the survivor function since the item "survives" for a time t. Mathematically, we can write the survivor function $R(t)$ as[1]

$$R(t) = 1 - F(t) \quad \text{for } t > 0 \tag{1.4}$$

Recall that $F(t)$ represents the probability that an item fails on the interval $(0, t]$ so logically that the reliability is simply one minus that probability. Figure 1.1 shows the familiar Gaussian CDF and the associated reliability function $R(t)$.

1.1.2 Availability

In the telecommunications environment, the metric most often used in contracts, designs, and discussion is availability. The dictionary defines available as "present or ready for immediate use." This definition has direct applicability in the world of telecommunications. When an item or a system is referred to as being "available," it is inherently implied that the system is working. When the item or system is referred to as "unavailable," it is implied that the system has failed. Thus, when applied to a telecommunications item or system, the term availability implies how ready a system is for use. The technical definition of availability (according to MIL-STD-721C) is:

"A measure of the degree to which an item or system is in an operable and committable state at the start of a mission when the mission is called for an unknown (random) time."

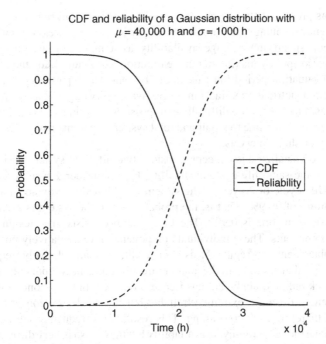

Figure 1.1. Gaussian CDF and associated reliability function $R(t)$.

Both the technical definition and the qualitative dictionary definition have the same fundamental meaning. This meaning can be captured by asking the question: "Is the system ready for use at any particular instant in time?" The answer to this question can clearly be formulated in terms of a statistical probability of readiness.

1.1.2.1 Availability Calculations. When examined as a statistical quantity, the availability of an item or a system can take on two different quantitative definitions. The average availability of an item or a system is the statistical probability of that item or system working over a defined period of time. For example, if an item's or a system's life cycle is considered to be 5 years and the availability of that item or system is of interest, then the availability of that item or system can be calculated as

$$A = \frac{\text{item or system uptime}}{\text{item or system operational time}} \tag{1.5}$$

In this case, the availability of the item or system is defined in terms of the percentage of time the item or system is working with respect to the amount of time the item or system has been in operation. (Note that the term "item" is used as shorthand to denote any item, system, or subsystem being analyzed.) This form of calculation of availability provides an average or mean availability over a specific period of time (defined by the operational time). One interesting item of note in this calculation is that the average availability as presented above provides very little insight with regard to the duration and/or frequency of outages that might occur, particularly in cases of long operational periods. When specifying average availability as a metric or design criteria, it is important to also specify maximum outage duration and failure frequency. Availability lifecycle or evaluation period must be carefully considered, particularly when availability is used as a metric for contract language. Availability targets that are achievable on an annual basis may be very difficult or impossible to achieve on monthly or even quarterly intervals. The time to repair and total system downtime have a great impact on availability over short intervals.

In order to visualize this concept, consider two different system designs, both of which achieve the same life-cycle availability. First, consider a system design with a replacement life cycle of 20 years. The system is designed to provide an average life-cycle availability of 99.9%. That is, the probability that the system is available at any particular instant in time is 0.999. The first system consists of a design with many redundant components. These individual components have a relatively poor reliability and need replacement on a regular basis. As a result, there are relatively frequent, short-duration outages that result from the dual failure of redundant components. This system is brought back online quickly, but has frequent outages. In the second system design, the components in use are extremely reliable but due to design constraints repair is difficult and therefore time consuming. This results in infrequent, long outages. Both systems achieve the same life-cycle availability but they do so in very different manners. The customer that uses the system in question would be well advised to understand both the mean repair time for a system failure as well as the most common expected failure modes in order to ensure that their expectations are met. Figure 1.2 provides a graphical

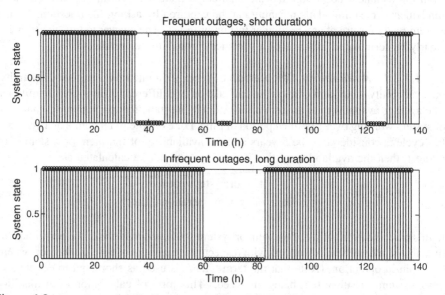

Figure 1.2. Average availability for system 1 (short duration, frequent outages) and system 2 (long duration, infrequent outages).

sketch of the scenario described above (note that the time scale has been exaggerated for emphasis).

The technical definition of availability need not be limited to the average or mean value. Availability can also be defined in terms of a time-dependent function $A(t)$ given by[1]

$$A(t) = Pr(X(t) = 1) \quad \text{for all } t \geq 0 \tag{1.6}$$

The term $A(t)$ specifies availability for a moment in time and is thus referred to as the instantaneous availability. The introduction of time dependence to the calculation of availability implies that the availability of an item can change with time. This could be due to a number of factors including early or late item failures, maintenance/repair practice changes, or sparing considerations. In most telecommunications system analyses, the steady-state availability is commonly used for system design or for contract language definitions. This assumption may not be appropriate for systems that require burn in or significant troubleshooting during system turn-up. Likewise, the system may become more unavailable as the system ages and vendors discontinue the manufacture of components or items begin to see late failures. The instantaneous availability $A(t)$ is related to the average availability A by the expression[1]

$$A_{\text{Average}} = \frac{1}{t_2 - t_1} \int_{t_1}^{t_2} A(t) dt \tag{1.7}$$

The most familiar form that availability takes in telecommunications system analysis is in relation to the mean time between failures (MTBF) and the mean time to repair (MTTR). These terms refer to the average (mean) amount of time that an item or a system is functioning (MTBF) between failure events and the average (mean) amount of time that it takes to place the item or system back into service. The average availability of a system can thus be determined by calculating[1]

$$A_{\text{Average}} = \frac{\text{MTBF}}{(\text{MTBF} + \text{MTTR})} \tag{1.8}$$

Availability is the average time between failures (operational time) divided by the average downtime plus the operational time (total time).

Unavailability is defined as the probability that the system is not functional at any particular instant in time or over a defined period of time. The expression for instantaneous unavailability is

$$U(t) = Pr(X(t) = 0) \quad \text{for all } t \geq 0 \tag{1.9}$$

where $U(t)$ represents time-dependent unavailability. The average value of unavailability is given by

$$U_{\text{Average}} = 1 - A_{\text{Average}} = \frac{\text{MTTR}}{(\text{MTBF} + \text{MTTR})} \tag{1.10}$$

It should be clear to the reader that calculations performed using the average expressions above are broad brush averages and do not give much insight into the variability of repair or failure events in the item or system. Calculation of availability using the expression above assumes that the sample set is very large and that the system achieves the average behavior. The applicability of this average availability value varies from system to system. In cases of relatively small-deployed component counts, this number may significantly misrepresent the actual achieved results of a system. For example, if the availability of a particular device (only one device is installed in the network of interest) is calculated using the average value based on a vendor provided MTBF and an assumed MTTR, one might be led to believe that the device availability is within the specifications desired. Consider a case where the MTTR has a high variability (statistical variance). Also consider that the device MTBF is very large, such that it might only be expected to fail once or twice in its lifetime. The achieved availability and the average availability could have very different values in this case since the variability of the repair period is high and the sample set is very small. The availability analyst must make careful consideration of not only the average system behavior but also the boundary behavior of the system being analyzed.

Bounding the achievable availability of an item or a system places bounds on the risk. Risk can be financial, technical, political, and so on, but risk is always present in a system design. Developing a clear understanding of system failure modes, expected system performance (both average and boundary value), and system cost reduces risk significantly and allows all parties involved to make the best, most informed decisions regarding construction and operations of a telecommunications system.

1.1.3 Maintainability

Maintainability as a metric is a measure of how quickly and efficiently a system can be repaired in order to ensure performance within the required specifications. MIL-STD-721C defines maintainability as:

> "The measure of the ability of an item to be retained in or restored to specified condition when maintenance is performed by personnel having specified skill levels, using prescribed procedures and resources, at each prescribed level of maintenance and repair."

The most common metric of maintainability used in telecommunications systems is the MTTR. This term refers to the average amount of time that a system is "down" or not operational. This restoral period can apply to either planned or unplanned outage events.

In the telecommunications environment, two types of downtime are typically tracked or observed. There are downtime events due to planned system maintenance such as preventative maintenance (PM), system upgrades, and system reconfiguration or growth. These types of events are typically coordinated with between the system operator and the customer and commonly fall outside of the contractual availability calculations. The second type of downtime event is the outage that occurs due to a failure in the system that results in a service outage. This system downtime is most commonly of primary interest to system designers, operators, and customers.

Scheduled or coordinated maintenance activities typically have predetermined downtime that are carefully controlled to ensure compliance with customer expectations. Such planned maintenance normally has shorter outage durations than unplanned maintenance or repair. Unplanned outages usually require additional time to detect the outage, diagnose its location, mobilize the repair activity, and get to the location of the failure to effect the repair. Unplanned outages that result from system failures result in downtime with varying durations. The duration and variability of the outage durations is dependent on the system's maintainability. A highly maintainable system will have a mean restoral period that is low relative to the system's interfailure period. In addition, the variance of the restoral period will also be small that ensures consistent, predictable outage durations in the case of a system failure event.

MTTR is commonly used interchangeably with the term mean downtime (MDT). MDT represents the sum of the MTTR and the time it takes to identify the failure and to dispatch for repair. Failure identification and dispatch in telecommunications systems can vary from minutes to hours depending on the system type and criticality.

In simple analyses, MDT is modeled assuming an exponential statistical distribution in which a repair rate is specified. Although this simplifying assumption makes the calculations more straightforward, it can result in significant inaccuracies in the resulting conclusions. Telecommunications system repairs more accurately follow normal or lognormal statistical distributions in which the repair of an item or a system has boundaries on the minimum and maximum values observed. The boundaries can be controlled by specifying both the mean and standard deviation of the repair period and by defining the distribution of repair based on those specifications.

MDT can be empirically calculated by collecting real repair data and applying best-fit statistical analysis to determine the distribution model and parameters that best represent the collected dataset.

1.1.4 Mean Time Between Failures, Failure Rates, and FITs

The most fundamental metric used in the analysis, definition, and design of telecommunications components is the MTBF. The MTBF is commonly specified by vendors and system engineers. It is a figure of merit describing the expected performance to be obtained by a component or a system. MTBF is typically provided in hours for telecommunications systems.

The failure rate metric is sometimes encountered in telecommunications systems. The failure rate describes the rate of failures (typically in failures per hour) as a function of time and in the general case is not a constant value. The most common visualization of failure rate is the bathtub curve where the early and late failure rates are much higher than the steady-state failure rate of a component (bottom of the bathtub). Figure 1.3 shows a sketch of the commonly observed "bathtub" curve for electronic systems. Note that although Figure 1.3 shows the failure rates early in system life and late in system life as being identical, in general, both the rate of failure rate change $dz(t)/dt$ and the initial and final values of failure rate are different.

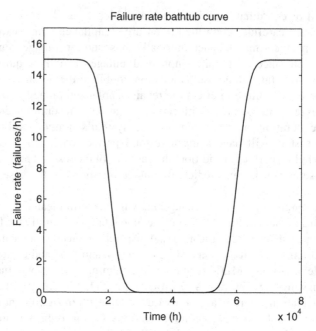

Figure 1.3. Bathtub curve for electronic systems.

A special case of the failure rate metric is the failures in time (FITs) metric. FITs are simply the failure rate of an item per billion hours:

$$\text{FITS} = \frac{z(t)}{10^9} \tag{1.11}$$

where $z(t)$ is the time-dependent failure rate expression. FITS values provided for telecommunications items are almost exclusively constant.

1.1.4.1 MTBF. The mean time to failure defines the average or more specifically the expected value of the TTF of an item, subsystem, or a system. Reliability and availability models rely upon the use of random variables to model component performance. The TTF of an item, subsystem or system is represented by a statistically distributed random variable. The MTTF is the mean value of this variable. In almost all telecommunications models (with the exception of software and firmware), it is assumed that the TTF of a component is exponentially distributed and thus the failure rate is constant (as will be shown in Section 1.2.1). The mean time to failure can be mathematically calculated by applying (Bain and Englehardt, 1992)

$$\text{MTTF} = E[\text{TTF}] = \int_0^\infty t \cdot f(t)dt \tag{1.12}$$

This definition is the familiar first moment or expected value of a random variable. The commonly used MTBF can be approximated by the MTTF when the repair or restoral time (MDT) is small with respect to the MTTF. Furthermore, if the MTTF $< \infty$, then we can write the MTTF as (by applying $f(t) = -R'(t))$[1]

$$\text{MTTF} = \int_0^\infty R(t)dt \tag{1.13}$$

This expression is particularly useful for calculating the MTTF (or MTBF) in many circumstances.

Telecommunications engineers must be particularly careful when using vendor-provided MTBF values. In many cases, the MTBF and the failure rate are presented as interchangeable inverses of each other. This special case is only true if one assumes that the TTF of a component is exponentially distributed. If the TTF of a component is not assumed to be exponentially distributed, this condition does not hold.

$$z(t) = -\frac{d}{dt}\ln R(t) \tag{1.14}$$

Note that except in the case where the TTF or TTR is exponentially distributed, the resultant failure rate is not constant. It is typically safe to assume that the MTBF and failure rate are inverses of each other if steady-state operation is assumed (see Figure 1.3). In the steady-state operation case, the failure rate is constant and the assumption of exponentially distributed TTFs holds. Early and late failure rates are time dependent and the exponential distribution assumption is invalid. Furthermore, if the system being considered employs redundancy, it does not necessarily hold that the redundant combination of components is exponentially distributed.

1.1.4.2 Failure Rates and FITs.
The mathematical definition of failure rate is the probability that an item fails on an infinitesimally small interval (Δt) given that it has not failed at time t[1]

$$Pr(t < T \le t + \Delta t \mid T > t) = \frac{Pr(t < T \le t + \Delta t)}{Pr(T > t)} = \frac{F(t + \Delta t) - F(t)}{R(t)} \tag{1.15}$$

If we take equation 1.15 and divide by an infinitesimally small time Δt (on both the LHS and RHS), then the failure rate $z(t)$ is given by[1]

$$z(t) = \lim_{\Delta t \to 0} \frac{F(t + \Delta t) - F(t)}{\Delta t} \cdot \frac{1}{R(t)} = \frac{f(t)}{R(t)} \tag{1.16}$$

The failure rate of an item or a component can be empirically determined by examining the histogram statistics of failure events. Empirical determination of the failure rate of a component in telecommunications can provide valuable information. It is therefore important to collect failure data in an organized, searchable format such as a database.

This allows post processors to determine time to failure and failure mode. Empirical failure rate determination is of particular value for systems where the deployed component count is relatively high (generally greater than approximately 25 items for failure rates observed in nonredundant telecommunications systems). In these cases, the system will begin to exhibit observable statistical behavior. Observation of these statistics allows the system operator or system user to identify and address systemic or recurring problems within the system.

The empirical failure rate of a system can be tabulated by separating the time interval of interest into k disjoint intervals of duration Δt. Let $n(k)$ be the number of components that fail on the kth interval and let $m(k)$ be the number of functioning components on the kth interval. The empirical failure rate is the number of failures per interval functioning time. Thus, if each interval duration is Δt[1]

$$z(k) = \frac{n(k)}{m(k) \cdot \Delta t} \tag{1.17}$$

In cases of a large number of deployed components, the calculation of empirical failure rate can allow engineers to validate assumptions about failure distributions and steady-state conditions. Continuous or ongoing calculations of empirical failure rate can allow operators to identify infant mortality conditions or wear out proactively and preemptively deal with these issues before they cause major service-affecting outages.

Typical telecommunications engineers commonly encounter failure rates and FITs values when specifying subsystems or components during the system design process. Failure rates are rarely specified by vendors as time-dependent values and must be carefully examined when used in reliability or availability analyses. The engineer must ask him or herself whether the component failure rate is constant from a practical standpoint. If the constant failure rate assumption is valid, the engineer must then apply any redundancy conditions or requirements to the analysis. As will be seen later in this book, analysis of redundant systems involves several complications and subtleties that must be considered in order to produce meaningful results.

1.2 STATISTICAL DISTRIBUTIONS

System reliability analysis relies heavily on the application of theories developed in the field of mathematical probability and statistics. In order to model the behavior of telecommunications systems, the system analyst must understand the fundamentals of probability and statistics and their implications to reliability theory. Telecommunications system and component models typically use a small subset of the modern statistical distribution library. These distributions form the basis for complex failure and repair models. This section presents the mathematical details of each distribution of interest and discusses the applications for which those models are most relevant. The last section discusses distributions that may be encountered or needed on rare occasions. Each distribution discussion presents the distribution probability density function (PDF) and cumulative distribution function (CDF) as well as the failure rate or repair rate of the distribution.

1.2.1 Exponential Distribution

The exponential distribution is a continuous statistical distribution used extensively in reliability modeling of telecommunications systems. In reliability engineering, the exponential distribution is used because of its memory-less property and its relatively accurate representation of electronic component time to failure.[1] As will be shown in Section 1.3, there are significant simplifications that can be made if a component's time to failure is assumed to exponential.

The PDF of the exponential distribution is given by (Bain and Englehardt, 1992)

$$f(x) = \begin{cases} \lambda e^{-\lambda x} & \text{for } x \geq 0 \\ 0 & \text{for } x < 0 \end{cases} \tag{1.18}$$

Figure 1.4 shows a plot of the exponential PDF for varying values of λ. The values of λ selected for the figure reflect failure rates of one failure every 1, 3, or 5 years. These selections are reasonable expectations for the field of telecommunications and depend upon the equipment type and configuration.

Recalling that the CDF (Figure 1.5) for the exponential distribution can be calculated from the PDF (Bain and Englehardt, 1992)

$$F(x) = \int_0^{\infty} f(x)dx = \begin{cases} 1 - e^{-\lambda x} & \text{for } x \geq 0 \\ 0 & \text{for } x < 0 \end{cases} \tag{1.19}$$

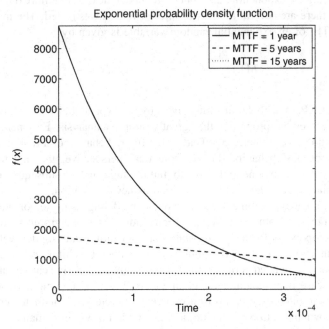

Figure 1.4. Exponential distribution PDF for varying values of λ.

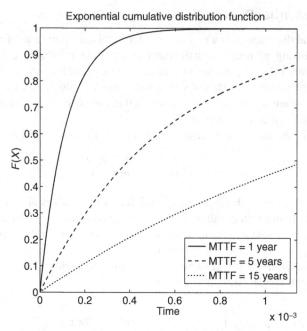

Figure 1.5. Exponential distribution CDF for varying values of λ.

Figure 1.5 plots the CDF for the same failure rates presented in Figure 1.4.

When using the exponential distribution to model the time to failure of an electronic component, there are several metrics of interest to be investigated. The mean time to failure (MTTF) of an exponential random variable is given by

$$\text{MTTF} = E[X] = \int_0^{\infty} x \cdot f(x)dx = \frac{1}{\lambda} \tag{1.20}$$

where $X \sim \text{EXP}(\lambda)$ with failure rate given by λ. Exponentially distributed random variables have several properties that greatly simplify analysis. Exponential random variables do not have a "memory." That is, the future behavior of a random variable is independent of past behavior. From a practical perspective, this means that if a component with an exponential time to failure fails and is subsequently repaired that repair places the component in "as good as new" condition.

The historical development of component modeling using exponential random variables is derived from the advent of semiconductors in electronic systems. Semiconductor components fit the steady-state constant failure rate model well. After an initial burn-in period exposes early failures, semiconductors exhibit a relatively constant failure rate for an extended period of time. This steady-state period can extend for many years in the case of semiconductor components. Early telecommunications systems consisted of circuit boards comprised of many discrete semiconductor components. As will be shown in Section 1.3, the failure rate of a serial system of many exponentially distributed semiconductor components is simply the sum of the individual component failure rates. Furthermore, since the sum of individual exponential random variables is an exponentially random variable, the failure rate of the resultant circuit board is exponentially distributed.

Modern telecommunications systems continue to use circuit boards comprised of many semiconductor devices. Modern systems use programmable components consisting of complex software modules. This software complicates analysis of telecommunications systems. Although the underlying components continue to exhibit exponentially distributed failure rates, the software operating on these systems is not necessarily exponentially distributed.

Although the exponential distribution is commonly used to model component repair, it is not well suited for this task. The repair of components typically is much more accurately modeled by normal, lognormal, or Weibull distributions. The reason that repair is typically modeled by an exponential random variable is due to the ease of analysis. As will be shown in Section 1.3, both the reliability block diagram (RBD) and Markov chain techniques of analysis rely upon the analyst assuming that repairs can be modeled by an exponential random variable. When the repair period of a system is very small with respect to the time between failures, this assumption is reasonable. When the repair period is not insignificant with respect to the time between failures, this assumption does not hold.

1.2.2 Normal and Lognormal Distributions

The normal (Gaussian) and lognormal distributions are continuous statistical distributions used to model a multitude of physical and abstract statistical systems. Both distributions can be used to model a large number of varying types of system repair behavior. In telecommunications systems, the failure can many times be well represented by the exponential distribution. Repair is more often well modeled by normal or lognormal random variables. System analysts or designers typically make assumptions or collect empirical data to support their system time to repair model selections. It is common to model the repair of a telecommunications system using a normal random variable since the normal distribution is completely defined by the mean and variance of that variable. These metrics are intuitive and useful when modeling system repair. In cases where empirical data is available, performing a best-fit statistical analysis to determine the best distribution for the time to repair model is recommended.

The PDF of the normal distribution (Bain and Englehardt, 1992) is given as

$$f(x, \mu, \sigma^2) = \frac{1}{\sqrt{2\pi\sigma^2}} e^{-\frac{(x-\mu)^2}{2\sigma^2}} \tag{1.21}$$

The normal distribution should be familiar to readers. The mean value (μ) represents the average value of the distribution while the standard deviation (σ) is a measure of the variability of the random variable. The lognormal distribution is simply the distribution of a random variable whose logarithm is normally distributed. Figure 1.6 shows the PDF of a normal random variable with $\mu = 8\,h$ and $\sigma = 2\,h$. These values of mean and standard deviation represent the time to repair for an arbitrary telecommunications system.

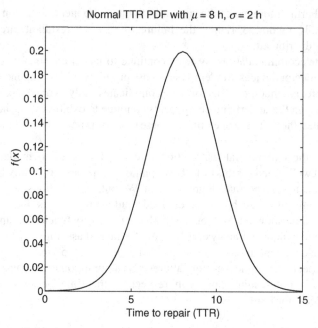

Figure 1.6. Normal distribution PDF of TTR, where $\mu = 8\,$h and $\sigma = 2\,$h.

The CDF of the normal distribution is given by a relatively complex expression involving the error function (erf) (Bain and Englehardt, 1992).

$$F(x,\ \mu, \sigma^2) = \frac{1}{2}\left(1 + \mathrm{erf}\left(\frac{x - \mu}{\sigma\sqrt{2}}\right)\right) \tag{1.22}$$

Figure 1.7 shows the cumulative distribution function of the random variable in Figure 1.6. The CDF provides insight into the expected behavior of the modeled repair time. The challenge in application of normally distributed repair models comes from the combination of these random variables with exponentially distributed failure models. Neither the reliability block diagram nor the Markov chain techniques allow the analyst to use any repair distribution but the exponential distribution. The most practical method for modeling system performance using normal, lognormal, or Weibull distributions is to apply Monte Carlo methods. Reliability and failure rate calculations are not presented in this section as it would be very unusual to use a normally distributed random variable to model the time to failure of a component in a telecommunications system. Exceptions to this might occur in submarine cable systems or wireless propagation models.

1.2.3 Weibull Distribution

The Weibull distribution is an extremely flexible distribution in the field of reliability engineering. The flexibility of the Weibull distribution comes from the ability to model many different lifetime behaviors by careful selection of the shape (α) and scale (λ)

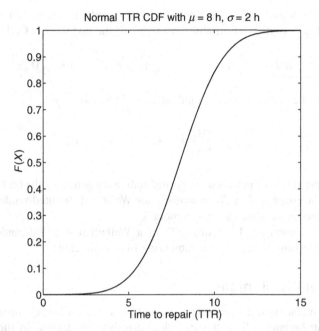

Figure 1.7. Normal distribution CDF of TTR, where $\mu = 8\,h$ and $\sigma = 2\,h$.

parameters. Generally, all but the most sophisticated telecommunications systems failure performance models use exponentially distributed time to failure. The Weibull distribution gives the analyst a powerful tool for modeling the time to failure or time to repair of nonelectronic system components (such as fiber-optic cables or generator sets). Parameter selection for Weibull distributed random variables requires expert knowledge of component performance or empirical data to ensure that the model properly reflects the desired parameter.

The PDF of a Weibull distributed random variable $T \sim \text{Weibull}(\alpha, \lambda)$ with $\alpha > 0$ and $\lambda > 0$ is given by equation 1.23 while the CDF of the time to failure T is given by equation 1.24 (Bain and Englehardt, 1992).

$$f(t) = \begin{cases} \alpha \lambda^{\alpha} t^{\alpha-1} e^{-(\lambda t)^{\alpha}} & \text{for } t > 0 \\ 0 & \text{otherwise} \end{cases} \tag{1.23}$$

$$F(t) = Pr(T \leq t) = \begin{cases} 1 - e^{-(\lambda t)^{\alpha}} & \text{for } t > 0 \\ 0 & \text{otherwise} \end{cases} \tag{1.24}$$

The two parameters in the Weibull distribution are known as the scale (λ) and the shape (α). When the shape parameter $\alpha = 1$, the Weibull distribution is equal to the familiar exponential distribution where λ mirrors the failure rate as discussed in Section 1.2.1.

The reliability function of a Weibull distributed random variable can be calculated by applying the definition of reliability in terms of the distribution CDF. That is

$$R(t) = 1 - F(t) = Pr(T \geq t) = e^{-(\lambda t)^{\alpha}} \quad \text{for } t > 0 \qquad (1.25)$$

Recalling that the failure rate of a random variable is given by

$$z(t) = \frac{f(t)}{R(t)} = \alpha \, \lambda^{\alpha} t^{\alpha-1} \quad \text{for } t > 0 \qquad (1.26)$$

Empirical curve fitting or parameter experimentation are generally the best methods for selection of the shape and scale parameters for Weibull distributed random variables applied to telecommunications system models.

Figure 1.8 shows the PDF and CDF of a Weibull distributed random variable representing the time to repair of a submarine fiber-optic cable.

1.2.4 Other Distributions

The field of mathematical probability and statistics defines a very large number of statistical distributions. All of the statistical distributions defined in literature have potential for use in system models. The difficulty is in relating distributions and their parameters to physical systems.

System analysts and engineers must rely on academic literature, research, and expert knowledge to guide distribution selection for system models. This book focuses

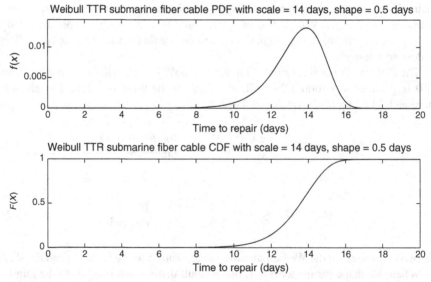

Figure 1.8. Weibull distributed random variable for submarine fiber-optic cable TTR.

on the presentation of relevant probability and statistics theory, and the concepts presented here have common practical application to telecommunications systems. More complex or less relevant statistical distributions not presented here are not necessarily irrelevant or inapplicable but rather must be used with care as they are not commonly used to model telecommunications systems.

On the most fundamental level, the entire behavior of a system or component is dictated by the analyst's selection of random variable distribution. As such, a significant amount of time and thought should be spent on the selection and definition of these statistical models. Care must be taken to ensure that the distribution selected is appropriate, relevant, and that it accurately reflects either the time to failure or time to repair behavior of the component of interest. Improper or incorrect distribution selection invalidates the entire model and the results produced by that model.

1.3   SYSTEM MODELING TECHNIQUES

Analysis of telecommunications systems requires accurate modeling in order to produce relevant, useful results. The metrics discussed in Section 1.1 are calculated by developing and analyzing system models. Many different reliability and availability modeling techniques exist. This book presents the methods and theories that are most relevant to the modeling and analysis of telecommunications systems. These techniques include RBD models, Markov chains, and Monte Carlo simulation. Each method has advantages and disadvantages. RBDs lend themselves to quick and easy results but sacrifice flexibility and accuracy, particularly when used with complex system topologies. Markov chain analysis provides higher accuracy but can be challenging to apply and requires models to use exponentially distributed random variables for both failure and repair rates. Monte Carlo simulation provides the ultimate in accuracy and flexibility but is the most complex and challenging to apply and is computationally intensive, even for modern computing platforms.

Availability is the most common metric analyzed in telecommunications systems design. Although reliability analysis can produce interesting and useful information, most systems are analyzed to determine the steady-state average (or mean) availability. RBDs and Markov chains presented in this chapter are limited to providing mean values of reliability or availability. Monte Carlo simulation techniques can be used to calculate instantaneous availabilities for components with nonconstant failure rates. The following sections present model theory and analysis techniques for each method discussed.

1.3.1   System Reliability

Analysis of system reliability requires the evaluation of interacting component random variables used to model failure performance of a system. This analysis is performed by evaluating the state of n discrete binary state variables $X_i(t)$, where $i = 1, 2, \ldots, n$.

Recall that the reliability function of a component is the probability that the component survives for a time t. Thus, the reliability of each component state variable $X_i(t)$ can be written as[1]

$$E[X_i(t)] = 0 \times Pr(X_i(t) = 0) + 1 \times Pr(X_i(t) = 1) = R_i(t) \text{ for } i = 1, 2, \ldots, n. \quad (1.27)$$

Equation 1.27 can be extended to the system case by applying[1]

$$R_S(t) = E[S(t)] \quad (1.28)$$

where $S(t)$ is the structure function of the component state vector $\mathbf{X}(t) = [X_1, X_2, \ldots, X_n]$. If we assume that the components of the system are statistically independent, then it can be shown that:[1]

$$R_S(t) = h(R_1(t), R_2(t), \ldots, R_n(t)) = h(R(t)) \quad (1.29)$$

1.3.2 Reliability Block Diagrams

RBDs are a common method for modeling the reliability of systems in which the order of component failure is not important and for which no repair of the system is considered. Many telecommunications engineers and analysts incorrectly apply parallel and serial reliability block diagram models to systems in which repair is central to the system's operation. Results obtained by applying RBD theory to availability models can produce varying degrees of inaccuracy in the output of the analysis. RBDs are success-based networks of components where the probability of mission success is calculated as a function of the component success probabilities. RBD theory can be understood most easily by considering the concept of a structure function. Figure 1.9 shows the reliability block diagram for both a series and a parallel combination of two components.

1.3.2.1 Structure Functions. Consider a system comprised of n independent components each having an operational state x_i. We can write the state of the ith component as shown in equation 1.30.[1] This analysis considers the component x_i to be a binary variable taking only one of two states (working or failed).

$$x_i = \begin{cases} 1 & \text{if the component is working} \\ 0 & \text{if the component has failed} \end{cases} \quad (1.30)$$

Thus, the system state vector $\mathbf{x}$ can be written as $\mathbf{x} = (x_1, x_2, x_3, \ldots, x_n)$. If we assume that knowledge of the individual states of x_i in $\mathbf{x}$ implies knowledge of the state of $\mathbf{x}$, we can write the structure function $S(\mathbf{x})$[1]

$$S(\mathbf{x}) = \begin{cases} 1 & \text{if the system is working} \\ 0 & \text{if the system has failed} \end{cases} \quad (1.31)$$

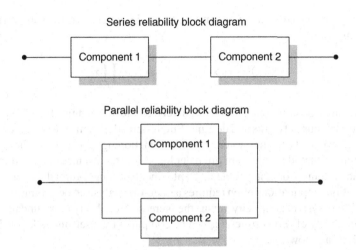

Figure 1.9. Series and parallel reliability block diagrams.[1]

where $S(\mathbf{x})$ is given by[1]

$$S(\mathbf{x}) = S(x_1, x_2, x_3 \ldots, x_n) \tag{1.32}$$

Thus, the structure function provides a resultant output state as a function of individual component states. It is important to note that reliability block diagrams are success-based network diagrams and are not always representative of system functionality. Careful development of RBDs requires the analyst to identify components and subsystems that can cause the structure function to take on the "working" or "failed" system state. In many cases, complex systems can be simplified by removing components from the analysis that are irrelevant. Irrelevant components are those that do not change the system state regardless of their failure condition.

RBDs can be decomposed into one of two different constituent structure types (series or parallel). It is instructive to analyze both of these system structures in order to develop an understanding of system performance and behavior. These RBD structures will form the basis for future reliability and availability analysis discussions.

1.3.2.2 Series Structures. Consider a system of components for which success is achieved if and only if all the components are working. This component configuration is referred to as a series structure (Figure 1.10). Consider a series combination of n components. The structure function for this series combination of

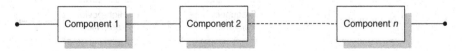

Figure 1.10. Series structure reliability block diagram.

components can be written as shown in equation 1.33, where x_n is the state variable for the nth component.[1]

$$S(\mathbf{x}) = x_1 \times x_2 \times \ldots \times x_n = \prod_{i=1}^{n} x_i \tag{1.33}$$

Series structures of components are often referred to as "single-thread" systems in telecommunications networks and designs. Single-thread systems are so named because all of the components in the system must be functioning in order for the system to function. Single-thread systems are often deployed in circumstances where redundancy is either not required or not practical. Deployment of single-thread systems in telecommunications applications often requires a trade-off analysis to determine the benefits of single-thread system simplicity versus the increased reliability of redundant systems.

The reliability of series structures can be computed by inserting equation 1.33 into equation 1.29 as shown below[1]

$$S(X(t)) = \prod_{i=1}^{n} X_i(t) \tag{1.34}$$

$$R(S(t)) = E\left[\prod_{i=1}^{n} X_i(t)\right] = \prod_{i=1}^{n} E[X_i(t)] = \prod_{i=1}^{n} R_i(t) \tag{1.35}$$

It is worth noting that the reliability of the system is at most as reliable as the least reliable component in the system[1]

$$R(S(t)) \leq \min(R_i(t)) \tag{1.36}$$

Figure 1.11 shows a single-thread satellite link RF chain and the reliability block diagram for that system. The reliability of the overall system is calculated below.

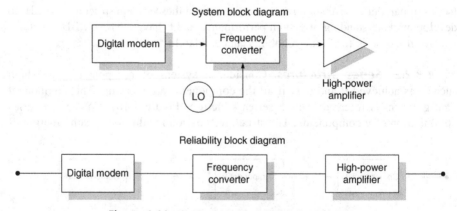

Figure 1.11. Single-thread satellite link RF chain.

Table 1.1. RF Chain Model Component Performance

Frequency Converter	Digital Modem	High-Power Amplifier
MTBF = 95,000 h	MTBF = 120,000 h	MTBF = 75,000 h
$R(4{,}380\,\text{h}) = 95.5\%$	$R(4{,}380\,\text{h}) = 96.4\%$	$R(4{,}380\,\text{h}) = 94.3\%$

Assume that the components of the RF chain have the following representative failure metrics (Table 1.1). We will calculate the probability that system survives 6 months of operation ($t = (365 \times 24/2) = 4380\,\text{h}$).

If we apply equation 1.35, we find that the system reliability is given by:

$$R(S(t)) = \prod_{i=1}^{n} R_i(t) = R_{\text{converter}} \times R_{\text{modem}} \times R_{\text{SSPA}} = 86.8\%$$

Although the relative reliabilities of the frequency converter, modem, and SSPA components are similar, the serial combination of the three elements results in a much lower predicted system reliability. Note that the frequency converter reliability includes the local oscillator.

1.3.2.3 Parallel Structures.

Consider a system of components for which success is achieved if any of the components in the system are working. This component configuration is referred to as a parallel structure as shown in Figure 1.12. Consider a parallel combination of n components. The structure function for this parallel combination of components can be written as shown in equation 1.37, where x_n is the state variable for the nth component.[1]

$$S(x) = 1 - (1 - x_1) \times (1 - x_2) \times \ldots (1 - x_n) = 1 - \prod_{i=1}^{n}(1 - x_i) \qquad (1.37)$$

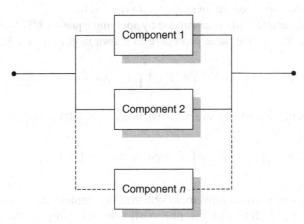

Figure 1.12. Parallel structure reliability block diagram.[1]

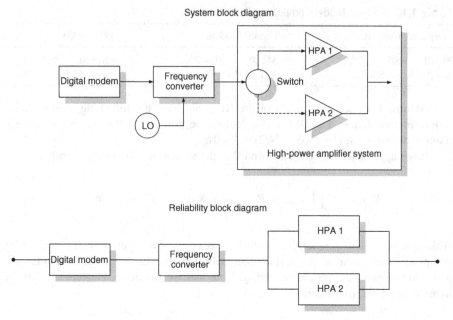

Figure 1.13. Parallel satellite RF chain system.

Parallel structures of components are often referred to as one-for-one or one-for-n redundant systems in telecommunications networks and designs. Redundant systems require the operation of only one of the components in the system for success. Figure 1.13 is a graphical depiction of a redundant version of the high power amplifier system portion of the satellite RF chain shown in Figure 1.11. This configuration of components dramatically increases the reliability of the RF chain but requires increased complexity for component failure switching. For the purposes of this simple example, the failure rate of the high-power amplifier switching component will be assumed to have a negligible impact on the overall system reliability.

Parallel system reliability is calculated by applying equation 1.37 to equation 1.29. The calculation follows the same procedure as shown in equations 1.34 and 1.35:

$$S(X(t)) = 1 - \prod_{i=1}^{n}(1 - X_i(t)) \qquad (1.38^1)$$

where $S(X(t))$ is the redundancy structure function for the HPA portion of the RF chain.

$$R(S(t)) = E\left[1 - \prod_{i=1}^{n}(1 - X_i(t))\right] = 1 - \prod_{i=1}^{n}(1 - R_i(t)) \qquad (1.39^1)$$

Examining the reliability improvement obtained by implementing redundant high-power amplifiers, we find that the previously low reliability of the single-thread amplifier now far exceeds that of the reliability of the single-thread modem and frequency converter. It can generally be found that the most significant improvement in system reliability performance can be obtained by adding redundancy to critical system components. Inclusion of secondary or tertiary redundancy systems continues to improve performance but does not provide the same initially dramatic increase in reliability that is observed by the addition of redundancy to a component or a subsystem.

$$R(S(t)) = R_{\text{converter}} \times R_{\text{modem}} \times R_{\text{HPA}}$$

In this case, R_{HPA} is a redundant system

$$R_{\text{SSPA}_{\text{System}}} = 1 - \prod_{i=1}^{2}(1 - R_{\text{SSPA}_i}) = 1 - (1 - R_{\text{SSPA}}) \times (1 - R_{\text{HPA}}) = 99.7\%$$

Thus, the total system reliability is now

$$R(S(t)) = R_{\text{converter}} \times R_{\text{modem}} \times R_{\text{HPA}_{\text{system}}} = 91.8\%$$

1.3.2.4 *k-Out-of-n Structures.*

The k-out-of-n structure is a system of components for which success is achieved if k or more of the n components in the system are working (this text assumes the "k-out-of-n: working" approach. A second approach is published in literature (Way and Ming, 2003), where success is achieved if k-out-of-n of the system components have failed "k-out-of-n: failed." This approach is not discussed here although the mathematics of this approach is very similar). This component configuration is referred to as a k-out-of-n structure. The parallel structure presented is a special case of the k-out-of-n structure, where $k = 1$ and $n = 2$ (one out of two). The structure function for this redundant combination of components can be written as shown in equation 1.40, where x_n is the state variable for the nth component (Way and Ming, 2003).

$$S(X) = \begin{cases} 1 & \text{if } \sum_{i=1}^{n} x_i \geq k \\ 0 & \text{if } \sum_{i=1}^{n} x_i < k \end{cases} \tag{1.40}$$

k-out-of-n structures occur commonly in telecommunications systems. They are implemented in multiplexer systems, power rectification and distribution and RF power amplifiers among other systems. The advantage of implementing a k-out-of-n redundancy structure is cost savings. For example, a one-for-two redundancy configuration has $k = 2$ and $n = 3$. The one-for-two redundancy configuration is common in solid state power amplifier systems and power rectification systems, where modularity allows for expansion and cost savings. In this configuration, one of the three modules is redundant and thus $k = 2$. The cost savings that are obtained in this configuration can be substantial.

Consider the system where parallel or one-for-one redundancy is implemented.

Total Modules Required $=$ (2 working modules) $+$ (2 protection modules) $=$ 4 modules

Now consider the one-for-two configuration.

Total Modules Required $=$ (2 working modules) $+$ (1 protection module) $=$ 3 modules

The trade-off in this configuration is cost versus failure performance. The one-for-one configuration represents a 33% increase in component count over the one-for-two system. As will be shown through system reliability analysis, the reduction in reliability is relatively small and is often determined to be a reasonable sacrifice.

Calculation of k-out-of-n system reliability can be performed by observing that since the component failure events are assumed to be independent, we find that summation of the component states $S(X)$ is a binomially distributed random variable[1]

$$S(X) = \sum_{i=1}^{n} X_i(t) \rightarrow S(X) \sim \text{bin}(n, R(t)) \tag{1.41}$$

Note this treatment assumes that all of the components in the redundant system are identical. Recalling the probability of a specific binomial combination event[1]

$$Pr(S(X) = y) = \binom{n}{y} R(t)^y (1 - R(t))^{n-y} \tag{1.42}$$

In the working k-out-of-n case, we are interested in the probability of the summation $S(X) \geq k$.[1]

$$Pr(S(X) \geq k) = \sum_{y=k}^{n} \binom{n}{y} R(t)^y (1 - R(t))^{n-y} \tag{1.43}$$

Equation 1.43 simply sums all of the discrete binomial probabilities for states in which the system is working.

Examination of the previously discussed HPA redundant system shows that the two-out-of-three configuration results in a relatively small reduction in reliability performance with a large cost savings (see Figure 1.14 for a system block diagram and the associated reliability block diagram for the 1:2 HPA system).

$$Pr(S(X) \geq 2) = \sum_{y=2}^{3} \binom{3}{y} R_{\text{HPA}}{}^y (1 - R_{\text{HPA}})^{3-y}$$

$$Pr(S(X) \geq 2) = \binom{3}{2} 0.943^2 (1 - 0.943)^1 + \binom{3}{3} 0.943^3 \approx 99.1\%$$

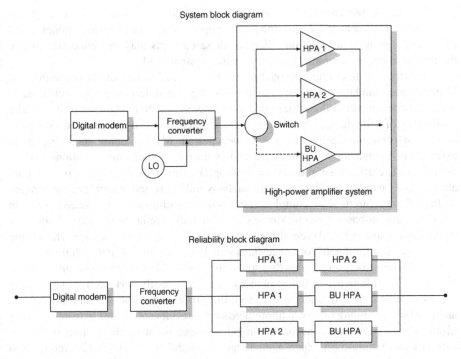

Figure 1.14. One-for-two (1:2) redundant HPA system block diagram.

1.4 SYSTEMS WITH REPAIR

The discussion of system failure performance up to this point has only examined systems in which repair is not possible or is not considered. Specifically, the term "reliability block diagram" refers to the success-based network of components from which system *reliability* is calculated. It is instructive to recall here that the definition of reliability is "the probability that an item can perform its intended function for a specified interval under stated conditions," as stated in Section 1.1. Thus, by definition, the behavior of the component or system following the first failure is not considered. In a reliability analysis, only the performance prior to the first failure is calculated.

For components (and systems of components), subject to repair after failure, different system modeling techniques must be used to obtain accurate estimates of system performance. The most common system performance metric used in repairable systems is availability. Availability is often used as a key performance indicator in telecommunications system design and frequently appears in contract service-level agreement (SLA) language. By specifying availability as the performance metric of interest, the analyst immediately implies that the system is repairable. Furthermore, by specifying availability, the applicability of RBDs as an analysis approach must be immediately discounted.

This section presents several key concepts related to repairable system analysis. These concepts include system modeling approaches, repair period models, and equipment sparing considerations. Each of these concepts plays an important role in the development of a complete and reasonable system model.

Two distinct modeling approaches are presented: Markov chain modeling and Monte Carlo simulation. Markov chain modeling is a state-based approach used to build a model in which the system occupies one of n discrete states at a time t. The probability of being in any one of the n states is calculated, thus resulting in a measure of system performance based on state occupation. Markov chain modeling is an extensively treated topic in literature and is useful in telecommunications system modeling of relatively simple system topologies. This book presents only a simple, abbreviated treatment of Markov chain analysis and interested readers are encouraged to do further research. More complex system configurations are better suited to Monte Carlo simulation-based models. Monte Carlo simulation refers to the use of numerical (typically computer-based) repetitive simulation of system performance. The Monte Carlo model is "simulated" for a specific system life many times and the lifetime failure statistics across many simulation "samples" are compiled to produce system performance statistics. Many performance metrics that can be easily derived from simulation results include failure frequency, time to failure (both mean and standard deviation), and availability among others. Although powerful results are available from applying Monte Carlo simulation, the development and execution of these models can be complex and tedious. An expert knowledge of reliability theory is often required to obtain confident results. Monte Carlo concepts and basic theory are presented in this section.

Repairable system models rely not only upon the assumed component TTF distributions (typically exponential) but also on the time to repair (TTR) distributions. It is shown in this chapter that one of the major drawbacks of applying Markov chain analysis techniques is that the TTR distribution must be exponential. This severely limits the models flexibility. In cases of electronic components or systems with TTF $\gg$ TTR, this assumption is often reasonable. It should be clear to the reader that an exponentially distributed random variable is an inherently poor model of telecommunications system repairs. Unfortunately, it is often the case in telecommunications systems that the TTF $\gg$ TTR assumption does not necessarily hold. This section presents the limitations and drawbacks of assuming an exponentially distributed TTR. In addition to the exponentially distributed time to repair, this section discusses Weibull, normal, and lognormal repair distributions and provides applications for these models.

The last section of this chapter presents the topic of system sparing. The concept of sparing in telecommunications systems should be familiar to anyone working in the field. Although the importance of sparing is typically recognized, it is often under-analyzed. Calculation of required sparing levels based on return material authorization (RMA) or component replacement period is presented. Cost implications and geographic considerations are also discussed. Component sparing can have significant impacts on the availability of a system but because it is not typically considered as part of the total system model, it is often overlooked and neglected.

1.5 MARKOV CHAIN MODELS

Consider a system consisting of a number of discrete *states* and *transitions* between those *states*. A Markov chain is a stochastic process (stochastic processes have behavior that is intrinsically nondeterministic) possessing the Markov property. The Markov property is simply the absence of "memory" within the process. This means that the current state of the system is the only state that has any influence on future events. All historical states are irrelevant and have no influence on future outcomes. For this reason, Markov processes are said to be "memory-less." It should be noted that Markov chains are not an appropriate choice for modeling systems where previous behavior has an affect on future performance.

To form a mathematical framework for the Markov chain, assume a process $\{X(t),\ t \geq 0\}$ with continuous time and a state space $\chi = \{0, 1, 2, \ldots, r\}$. The state of the process at a time s is given by $X(s) = i$, where i is the ith state in state space χ. The probability that the process will be in a state j at time $t + s$ is given by[1]

$$Pr(X(t + s) = j \mid X(s) = i),\ X(u) = x(u), 0 \leq u < s) \qquad (1.44)$$

where $\{x(u), 0 \leq u < s\}$ denotes the processes "history" up to time s. The process is said to possess the Markov property if[1]

$$\begin{aligned} Pr(X(t + s) = j \mid X(s) = i),\ X(u) = x(u), \\ 0 \leq u < s) = Pr(X(t + s) = j \mid X(s) = i) \quad \text{for all } x(u),\ 0 \leq u < s \end{aligned} \qquad (1.45)$$

Processes possessing the behavior shown in equation 1.45 are referred to as Markov processes. The Markov process treatment presented in this book assumes time-homogeneous behavior. This means that system global time does not affect the probability of transition between any two states i and j. Thus[1]

$$Pr(X(t + s) = j \mid X(s) = i) = Pr(X(t) = j \mid X(0) = i) \quad \text{for all } s, t \qquad (1.46)$$

Stated simply, equation 1.46 indicates that the probability of moving between states i and j is not affected by the current elapsed time. All moments in time result in the same probability of transition.

One classic telecommunications system problem is the calculation of availability for the one-for-one redundant system. Several different operational models exist in the one-for-one redundant system design. The system can be designed for hot-standby, cold-standby, or load-sharing operation. Each of these system design choices has an impact on the achievable system availability and the maintainability of the system. The Markov chain modeling technique is well suited to model systems of this type as long as the repair period is much shorter than the interfailure period (time to failure). This redundancy problem will be used to demonstrate the application and use of Markov chains in system modeling for the remainder of this section.

1.5.1   Markov Processes

Assume that a system can be modeled by a Markov process $\{X(t), t \geq 0\}$ with state space $\chi = \{0, 1, 2, \ldots, r\}$. Recall that the probability of transition between any two states i and j is time independent (stationary). The probability of a state transition from i to j is given by[1]

$$P_{ij}(t) = Pr(X(t) = j | X(0) = i) \quad \text{for all } i, j \in \chi \tag{1.47}$$

That is, P_{ij} is the probability of being in state j given that the system is in state i at time $t = 0$. One of the most powerful implications of the Markov process technique is the ability to represent these state transition probabilities in matrix form

$$P(t) = \begin{pmatrix} P_{00}(t) & \cdots & P_{0r}(t) \\ P_{10}(t) & & P_{1r}(t) \\ \vdots & \ddots & \vdots \\ P_{r0}(t) & \cdots & P_{rr}(t) \end{pmatrix} \tag{1.48}$$

Since the set of possible states $\chi = \{0, 1, 2, \ldots, r\}$ is finite and $i, j \in \chi$, for all $t \geq 0$, we find that the sum of all matrix row transition probabilities must necessarily be equal to unity.

$$\sum_{j=0}^{r} P_{ij}(t) = 1 \quad \text{for all } i \in \chi \tag{1.49}$$

The rows in the transition matrix represent the probability of a transition *out of* state i (where $i \neq j$) while the columns of the matrix represent the probability of transition *into* state j (where $i \neq j$).

From a practical perspective, the definition of a model using the Markov chain theory is relatively straightforward and simple. The approach presented here forgoes a number of mathematical subtleties in the interest of practical clarity. Readers interested in a more mathematical (and rigorous) treatment of the Markov chain topic are referred to Rausand and Høyland (2004).

As shown in equation 1.48, the Markov chain can be represented as a matrix of values indicating the probability of either entering or leaving a specific state in the state space χ. We introduce the term "sojourn time" to indicate the amount of time spent in any particular state i. It can be shown that the mean sojourn time in state i can be expressed as[1]

$$E(\tilde{T}_i) = \frac{1}{\alpha_i} \tag{1.50}$$

where α_i is the rate of transition from state i to another state in the state space (rate out of state i). Since the process is a Markov chain, it can also be shown that the sojourn time (and thus the transition rate α_i) must be exponentially distributed and that all sojourn times must be independent. These conditions ensure that the Markov chain's memory-less property is maintained. Analyses presented here assume $0 \leq \alpha_i \leq \infty$.

This assumption implies that no states are instantaneous ($\alpha_i \to \infty$) or absorbing ($\alpha_i \to 0$). Instantaneous states have a sojourn time equal to zero while absorbing states have an infinite sojourn time. We only consider states with finite sojourn durations.

Let the variable a_{ij} be the rate at which the process leaves state i and enters state j. Thus, the variable a_{ij} is the transition rate from i to j.[1]

$$a_{ij} = \alpha_i \times P_{ij} \quad \text{for all } i \neq j \tag{1.51}$$

Recall that α_i is the rate of transition out of state i and P_{ij} is the probability that the process enters state j after exiting state i. It is intuitive that when leaving state i, the process must fall into one of the r available states, thus[1]

$$\alpha_i = \sum_{\substack{j=0 \\ j \neq i}}^{r} a_{ij} \tag{1.52}$$

Since the coefficients a_{ij} can be calculated for each element in a matrix $\mathbb{A}$ by applying equation 1.51, we can define the transition rate matrix as shown in equation 1.53.[1]

$$\mathbb{A} = \begin{pmatrix} a_{00} & \cdots & a_{0r} \\ a_{10} & & a_{1r} \\ \vdots & \ddots & \vdots \\ a_{r0} & \cdots & a_{rr} \end{pmatrix} \tag{1.53}$$

The sum of all transition probabilities P_{ij} for each row must be equal to one, thus we can write the diagonal elements of $\mathbb{A}$ as[1]

$$a_{ii} = -\alpha_i = - \sum_{\substack{j=0 \\ j \neq i}}^{r} a_{ij} \tag{1.54}$$

The diagonal elements of $\mathbb{A}$ represent the sum of the departure and arrival rates for a state i. Markov processes can be visualized using a state transition diagram. This diagram provides an intuitive method for developing the transition rate matrix for a system model. It is common in state transition diagrams to represent system states by circles and transitions between states as directed segments. Figure 1.15 shows a state transition diagram for a one-for-one redundant component configuration. If both of the redundant components in the system are identical, the transition diagram can be further simplified (see Figure 1.16).

The procedure for establishing a Markov chain model transition rate matrix $\mathbb{A}$ involves several steps.

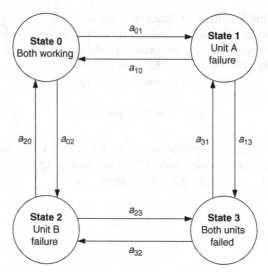

Figure 1.15. Redundant Markov chain state diagram.

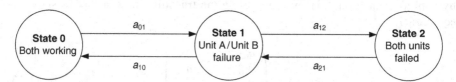

Figure 1.16. Redundant Markov chain state diagram, identical components.

Step 1. The first step in developing the transition rate matrix is to identify and describe all of the system states relevant to operation. Recall that relevant states are those that can affect the operation of the system. Irrelevant states are system states that do not affect system operation or failure, regardless of the condition. The identified relevant system states are then given an integer state identifier

$$S_i \in \chi, \quad \text{where } \chi = \{0, 1, \ldots, r\} \tag{1.55}$$

Step 2. Having identified the system states to be modeled, the transition rates to and from each state must be determined. In basic reliability analysis, these transition rates will almost always correspond to a component failure or repair rate. Component failure rates can typically be derived from system documentation, empirical data, or expertise in a field of study. Component repair rates are often based on assumptions, experience, or system require-ments. In Figure 1.15, the transition rates $\{a_{01}, a_{02}, a_{13}, a_{23}\}$ all represent component failure transition rates while the rates $\{a_{10}, a_{20}, a_{32}, a_{31}\}$ represent repair transition rates. Table 1.2 shows a tabulation of the transition rate, the common nomenclature used to represent each rate, and representative values for a 1:1 redundant high-power amplifier system.

Table 1.2. Markov Chain Transition Rate Matrix Table Example

Transition Rate	Commonly Used Term	Example Value (Failures/h)
$a_{01}, a_{02}, a_{13}, a_{23}$	λ_{HPA}	1.33×10^{-5}
$a_{10}, a_{20}, a_{32}, a_{31}$	μ_{HPA}	8.33×10^{-2}

Step 3. The values in Table 1.2 are inserted into the transition rate matrix $\mathbb{A}$ in their appropriate positions as shown below.

$$\mathbb{A} = \begin{pmatrix} a_{00} & a_{01} & a_{02} & a_{03} \\ a_{10} & a_{11} & a_{12} & a_{13} \\ a_{20} & a_{21} & a_{22} & a_{23} \\ a_{30} & a_{31} & a_{32} & a_{33} \end{pmatrix}$$

The tabulated failure and repair rates for each transition replace the interstate coefficients.

$$\mathbb{A} = \begin{pmatrix} a_{00} & \lambda_{HPA} & \lambda_{HPA} & 0 \\ \mu_{HPA} & a_{11} & 0 & \lambda_{HPA} \\ \mu_{HPA} & 0 & a_{22} & \lambda_{HPA} \\ 0 & \mu_{HPA} & \mu_{HPA} & a_{33} \end{pmatrix}$$

Step 4. The diagonal elements of the transition rate matrix are populated by applying equation 1.54 along each row. The resultant, completed transition rate matrix is shown below.

$$\mathbb{A} = \begin{pmatrix} -(\lambda_{HPA} + \lambda_{HPA}) & \lambda_{HPA} & \lambda_{HPA} & 0 \\ \mu_{HPA} & -(\lambda_{HPA} + \mu_{HPA}) & 0 & \lambda_{HPA} \\ \mu_{HPA} & 0 & -(\lambda_{HPA} + \mu_{HPA}) & \lambda_{HPA} \\ 0 & \mu_{HPA} & \mu_{HPA} & -(\mu_{HPA} + \mu_{HPA}) \end{pmatrix}$$

Careful consideration of the relevant states in Step 1 of the transition rate matrix definition can result in simplifications. Consider the system diagram shown in Figure 1.16. If the redundant components shown were assumed to be identical (as presented in Table 1.2), the system model could be shown as having three distinct states instead of four. The transition rate matrix for Markov chain in Figure 1.16 is given by

$$\mathbb{A} = \begin{pmatrix} -2\lambda_{HPA} & 2\lambda_{HPA} & 0 \\ \mu_{HPA} & -(\mu_{HPA} + \lambda_{HPA}) & \lambda_{HPA} \\ 0 & \mu_{HPA} & -\mu_{HPA} \end{pmatrix}$$

As Figure 1.16 shows, the complexity of the system model is greatly reduced with no loss of accuracy in the case where the two redundant components are identical.

1.5.2 State Equations

In order to solve the Markov chain for the relative probabilities of occupation for each system state, we must apply two sets of equations. Through analysis of the Chapman-Kolmogorov equations, it can be shown[1] that the following differential equation can be derived.

$$\dot{\mathbb{P}}(t) = \mathbb{P}(t) \cdot \mathbb{A} \qquad (1.56)$$

where $\mathbb{P}(t)$ is the time-dependent state transition probability matrix and $\mathbb{A}$ is the transition rate matrix. The set of equations resulting from the matrix in Equation 1.56 are referred to as the Kolmogorov forward equations.

Assuming that the Markov chain is defined to occupy state 0 at time $t = 0$, $X(0) = i$ and $P_i(0) = 1$ while all other probabilities $P_k(0) = 0$ for $k \neq i$. This simply means that by defining the system to start in state i at time $t = 0$, we have forced the probability of occupation for state i at time $t = 0$ to be unity while the probability of being in any other state is zero. By defining the starting state, we can simplify equation 1.56 to the following form.

$$\begin{pmatrix} a_{00} & \cdots & a_{0r} \\ a_{10} & & a_{1r} \\ \vdots & \ddots & \vdots \\ a_{r0} & \cdots & a_{rr} \end{pmatrix} \cdot \begin{bmatrix} P_0(t) \\ P_1(t) \\ \vdots \\ P_r(t) \end{bmatrix} = \begin{bmatrix} \dot{P}_0(t) \\ \dot{P}_1(t) \\ \vdots \\ \dot{P}_r(t) \end{bmatrix} \qquad (1.57)$$

Equation 1.57 does not have a unique solution but by applying the initial condition $(P_i(0) = 1)$ and recalling that the sum of each column is equal to one, we can often find a solution to the set of equations. In practical problems, it is rare that the system of equations does not result in a real, finite solution.

Solutions to equation 1.57 are time dependent. Analyses performed on telecommunications systems are often interested in the steady-state solution to equation 1.57. In these circumstances, we can further simplify our problem by examining the behavior of equation 1.57 as $t \to \infty$. It can be shown[1] that after a long time ($t \to \infty$), the probability of occupation for a particular system state is not dependent on the initial system state. Furthermore, if the probability of state occupation is constant, it is clear that the derivative of that probability is necessarily zero.

$$\lim_{t \to \infty} P_j(t) = P_j \quad \text{for } j = 1, 2, \ldots, r \qquad (1.58)$$

$$\lim_{t \to \infty} \dot{P}_j(t) = 0 \quad \text{for } j = 1, 2, \ldots, r \qquad (1.59)$$

Thus, we can rewrite equation 1.57 as

$$\begin{pmatrix} a_{00} & \cdots & a_{0r} \\ a_{10} & & a_{1r} \\ \vdots & \ddots & \vdots \\ a_{r0} & \cdots & a_{rr} \end{pmatrix} \cdot \begin{bmatrix} P_0 \\ P_1 \\ \vdots \\ P_r \end{bmatrix} = \begin{bmatrix} 0 \\ 0 \\ \vdots \\ 0 \end{bmatrix} \qquad (1.60)$$

Solution of equation 1.60 for each P_j relies upon use of linear set of algebraic equations and the column sum for each column j.

$$\sum_{j=0}^{r} P_j = 1 \tag{1.61}$$

1.5.3 State Equation Availability Solution

The system *availability* or *unavailability* is easily calculated once the state equations have been solved for the vector $\mathbb{P}$.

Define the set of all possible system states $S = \{S_0, S_1, \ldots, S_r\}$. Define a set W as the subset of S containing only the states in S where the system is working. Define another set F as the subset of S containing only those states where the system has failed. The availability of the system is the sum of all probabilities in W.[1]

$$A = \sum_{j \in W} P_j, \quad \text{where } W \in S \tag{1.62}$$

The unavailability of the system is likewise the sum of P_j over all states where the system has failed. Alternatively, the unavailability can be calculated by recognizing that the sum of the availability and unavailability must be unity.

$$1 = \sum_{j \in W} P_j + \sum_{j \in F} P_j \tag{1.63}$$

Replacing the sum in equation 1.63 and rearranging[1]

$$1 - A = \sum_{j \in F} P_j \tag{1.64}$$

Thus, calculating the availability immediately provides us with the unavailability as well.

1.6 PRACTICAL MARKOV SYSTEM MODELS

Markov system models have been used extensively in many industries to model the reliability of a variety of systems. Within the field of telecommunications and with the advent of modern computing techniques, the application of Markov chain modeling methods in telecommunications systems models is limited to a few special cases.

Markov models can provide quick, accurate assessments of redundant system availabilities for relatively simple topologies. Systems in which the time-to-repair distribution is not exponential or where the redundancy configuration is complex are not good candidates for practical Markov models. Complex mathematics and sophisticated matrix operations in those types of models should lead the engineer to consider Monte Carlo simulation in those circumstances.

The Markov chain modeling technique is well suited to redundancy models consisting of a small number of components. The mathematics of analyzing 1:1 or

1:2 redundancies remains manageable and typically do not require numerical computation or computer assistance. For this reason, the engineer can usually obtain results much more quickly than would be possible using a Monte Carlo analysis approach. In many cases, a full-blown system model is not required and only general guidelines are desired in the decision-making process.

Although the scope of practical Markov system models is somewhat limited, the types of problems that are well suited for Markov analysis are common and practical. This section will present the Markov model for the following system types.

1. Single-component system model
2. Hot-standby redundant system model
3. Cold-standby redundant system model

Each of the models listed above represent a common configuration deployed in modern telecommunications systems. These models apply to power systems, multiplexing systems, amplifier systems, and so on.

1.6.1 Single-Component System Model

The simplest Markov chain model is the model for a single component. This model consists of two system states $S = \{S_0, S_1\}$.

Let S_0 be the working component state and S_1 be the failed component state. Figure 1.17 is the Markov state transition diagram for this system model.

The transition rate matrix is very straightforward, consisting of four coefficients. If we let the failure rate of a component be defined as λ and the repair rate of the component be defined as μ, we have (by applying the steps listed previously)

$$\mathbb{A} = \begin{pmatrix} a_{00} & a_{01} \\ a_{10} & a_{11} \end{pmatrix} = \begin{pmatrix} -\lambda & \lambda \\ \mu & -\mu \end{pmatrix}$$

Applying equation 1.60, we can solve the state equations to determine the probabilities of state occupation $\mathbb{P} = [P_0 \ P_1]$.

$$\mathbb{P} \cdot \mathbb{A} = \bar{0} = [P_0 \ P_1] \cdot \begin{pmatrix} -\lambda & \lambda \\ \mu & -\mu \end{pmatrix}$$

The set of linear equations is thus:

$$-\lambda P_0 + \mu P_1 = 0 \quad (1)$$
$$\lambda P_0 - \mu P_1 = 0 \quad (2)$$
$$P_0 + P_1 = 1 \quad (3)$$

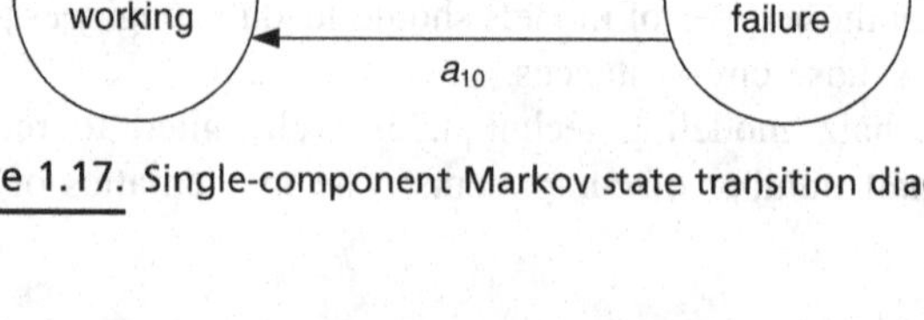

Figure 1.17. Single-component Markov state transition diagram.

Solving the equations using (1) and (3), we obtain

$$P_0 = \frac{\mu}{\mu + \lambda} \quad \text{and} \quad P_1 = \frac{\lambda}{\mu + \lambda}$$

Calculation of the component availability is straightforward once the individual state probabilities have been determined. Let the set $W = \{S_0\}$ and $F = \{S_1\}$. Thus, the availability of the system is simply equal to

$$A = \sum_{j \in W} P_j = P_0$$

Recall that we earlier made the assumption (in order to preserve the Markov property) that the state transition probabilities were exponentially distributed random variables and thus the transition rates λ and μ are constant, so we can write

$$\text{MTBF} = \frac{1}{\lambda}, \quad \text{MTTR} = \frac{1}{\mu}$$

If we rewrite the expression for P_0 in terms of MTBF and MTTR, we find

$$P_0 = \frac{\frac{1}{\text{MTTR}}}{\frac{1}{\text{MTTR}} + \frac{1}{\text{MTBF}}} = \frac{\text{MTBF}}{\text{MTBF} + \text{MTTR}}$$

This is the same result that was obtained in Section 1.1.

1.6.2 Hot-Standby Redundant System Model

Consider a system consisting of two identical components that are both operating continuously. This particular system does not implement load sharing but rather one of the two components carries the entire load at any given time. Upon failure of one of the components, the system immediately switches from the primary module to the backup (redundant) module.

In our hot-standby model (Figure 1.18), we have three system states $S = \{S_0, S_1, S_2\}$. Define the systems states as described in Table 1.3.

One of the disadvantages of the hot-standby redundancy configuration is that during operation, the backup module accumulates life-cycle operational hours that ultimately lead to the failure of that module. The module is in operation only to ensure that the system continues to operate if the primary module fails. In a cold-standby system, the backup module is not operated until such time that the primary module fails. This "saves" the operational hours of the backup modules for use when the component is doing real work.

Definition of the transition rate matrix follows the same procedure used previously.

$$\mathbb{A} = \begin{pmatrix} -2\lambda & 2\lambda & 0 \\ \mu & -(\mu + \lambda) & \lambda \\ \mu & 0 & -\mu \end{pmatrix}$$

Table 1.3. Hot-Standby System State Descriptions

State	System Operating Condition	Description
S_0	Working	Both modules working, system in nominal condition
S_1	Working	Single module failure, system operating with one module failed
S_2	Failure	Dual module failure, system failure

Applying the state equation matrix definition to determine the linear algebraic equations in terms of state occupation probabilities, $\mathbb{P} = [P_0 \; P_1 \; P_2]$.

$$\mathbb{P} \cdot \mathbb{A} = \bar{0} = \begin{bmatrix} P_0 & P_1 & P_2 \end{bmatrix} \cdot \begin{pmatrix} -2\lambda & 2\lambda & 0 \\ \mu & -(\mu + \lambda) & \lambda \\ \mu & 0 & -\mu \end{pmatrix}$$

$$-2\lambda P_0 + \mu P_1 + \mu P_2 = 0$$
$$2\lambda P_0 - (\mu + \lambda)P_1 = 0$$
$$\lambda P_1 - \mu P_2 = 0$$
$$P_0 + P_1 + P_2 = 1$$

Solving the simultaneous equations for $\mathbb{P}$, we find

$$P_0 = \frac{\mu}{2\lambda + \mu}$$
$$P_1 = \frac{2\lambda\mu}{(\lambda + \mu)(2\lambda + \mu)}$$
$$P_2 = \frac{2\lambda^2}{(\lambda + \mu)(2\lambda + \mu)}$$

We now define the subsets of S for which the system is working and failed. In the working case, we have $W = \{S_0, S_1\}$ and for the failed case, we have $F = \{S_2\}$. Thus, we

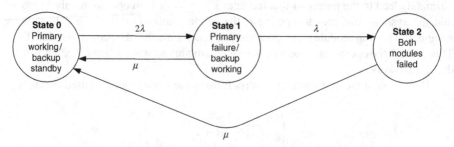

Figure 1.18. Hot-standby redundant Markov state transition diagram.

can calculate the availability of the system to be

$$A = \sum_{j \in W} P_j = P_0 + P_1 = \frac{\mu^2 + 3\lambda\mu}{(\lambda + \mu)(2\lambda + \mu)}$$

1.6.3 Cold-Standby Redundant Model

Analysis of the cold-standby redundant model follows the same process that was used in the hot-standby model. In this case, the assumptions are changed slightly, resulting in a modified state transition diagram and a different overall result. The diagram shown in Figure 1.19 shows the modified state transition diagram.

Note that in this case, we have assumed that a failure of both units will force a repair that places the working module back into operation and simultaneously repairs the standby module making it ready for service once again. Also note that during normal operation, only one of the two modules is accumulating operational hours ($a_{01} = \lambda$).

Continuing with the same analysis procedure used in the hot-standby case, we define each of the system states $S = \{S_0, S_1, S_2\}$ as in Table 1.4.

The transition rate matrix is given by

$$\mathbb{A} = \begin{pmatrix} -\lambda & \lambda & 0 \\ \mu & -(\mu + \lambda) & \lambda \\ \mu & 0 & -\mu \end{pmatrix}$$

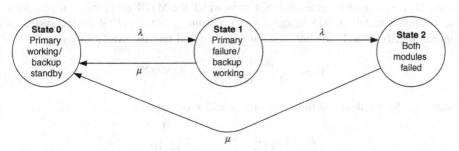

Figure 1.19. Cold-standby Markov state transition diagram.

Table 1.4. Cold-Standby System State Descriptions

State	System Operating Condition	Description
S_0	Primary working Backup standby	Primary module working, backup in standby mode, system working
S_1	Primary failed Backup working	Primary module failure, backup module operating, system working
S_2	Primary failed Backup failed	Primary module failure, backup module failure, system failure

Applying the state equation matrix definition to determine the linear algebraic equations in terms of state occupation probabilities, $\mathbb{P} = [P_0 \; P_1 \; P_2]$.

$$\mathbb{P} \cdot \mathbb{A} = \bar{0} = [P_0 \quad P_1 \quad P_2] \cdot \begin{pmatrix} -\lambda & \lambda & 0 \\ \mu & -(\mu + \lambda) & \lambda \\ \mu & 0 & -\mu \end{pmatrix}$$

Solving the simultaneous equations for $\mathbb{P}$, we find

$$P_0 = \frac{\mu}{\lambda + \mu}$$

$$P_1 = \frac{\lambda\mu}{\lambda^2 + 2\lambda\mu + \mu^2}$$

$$P_2 = \frac{\lambda^2}{\lambda^2 + 2\lambda\mu + \mu^2}$$

As previously mentioned, the availability is calculated by defining the subsets of S for which the system is working and failed. In the working case, we have $W = \{S_0, S_1\}$ and for the failed case, we have $F = \{S_2\}$. Thus, we can calculate the availability of the system to be

$$A = \sum_{j \in W} P_j = P_0 + P_1 = \frac{\mu^2 + 2\lambda\mu}{\lambda^2 + 2\lambda\mu + \mu^2}$$

As a comparison of relative performance between the hot-standby and cold-standby availabilities, consider a generator system in which the MTBF of a particular generator set is approximately 8000 h (about 1 year). Assume that the MTTR for the system is approximately 24 h. In the hot-standby case, we find that the availability is

$$A_{\text{hot}} = \frac{\mu^2 + 3\lambda\mu}{2\lambda^2 + 3\lambda\mu + \mu^2} \approx 99.9982\%$$

where we have calculated the values for μ and λ by applying

$$\mu = \frac{1}{\text{MTTR}} \quad \text{and} \quad \lambda = \frac{1}{\text{MTBF}}$$

The cold-standby case provides an increase in availability performance since the standby component is not operational until the primary unit fails. Even when the backup unit is called upon to operate, its time in service is very short compared with the primary unit.

$$A_{\text{cold}} = \frac{\mu^2 + 2\lambda\mu}{\lambda^2 + 2\lambda\mu + \mu^2} \approx 99.9991\%$$

It should be noted that in electronic telecommunications systems, the MTBF is generally very large (typically greater than 150,000 h) and the MTTR is often less than 8 h. It should be clear that the cold-standby redundancy configuration is preferable, particularly in systems where the failure rate is significantly increased in hot-standby operation. Within the hot-standby redundancy configuration is an approach known as monitored hot standby (MHSB). MHSB systems are often preferred when component or system MTBF values are large because the operator has confidence that the backup system will be operational when called upon (because it is monitored and has been known to be operational). A cold standby may not operate and may in fact fail when called upon suddenly to operate (e.g., when say a high-voltage supply power is applied), particularly after long durations without in-service testing. Additionally, cold standby systems may have a "warm-up" time, and thus may not provide uninterrupted service.

1.7 MONTE CARLO SIMULATION MODELS

All of the models presented thus far have assumed that both the TTF and the TTR of an item or a system follow an exponential distribution. The exponentially distributed random variable assumption lends mathematical simplicity to both the reliability block diagram and the Markov chain models. In both cases, the mathematics of analysis is sufficiently simple that quick results are possible. The results obtained are often useful for *what if* analyses and for small system designs.

Unfortunately, the limitations imposed by assuming exponentially distributed time to failure and time to repair for a system can lead to unrealistic or inaccurate results in many telecommunications systems. It is in these cases that Monte Carlo simulation is beneficial. Most system models produced using Monte Carlo simulation involve many hours of model development and implementation. Engineers considering use of Monte Carlo simulation for reliability/availability analysis on a particular project should consider the following questions in order to determine whether Monte Carlo simulation is the best fit.

1. What is the purpose of the analysis?
2. What is the budget of the project? Can it support the labor costs associated with the Monte Carlo simulation approach?
3. What is the expertise of the analyst and project team? Will there be sufficient knowledge to derive the maximum benefit from a sophisticated analysis?
4. What are the specific reasons that reliability block diagrams and/or Markov chain analysis are not sufficient to meet the analysis requirements?
5. Does the project involve repairs that are not reasonably approximated by an exponentially distributed random variable?

In many cases, it is possible to make simplifying assumptions in the system model that allow reasonable results to be obtained without embarking on a full Monte Carlo system simulation. This section discusses the theory of Monte Carlo simulation.

1.7.1 System Modeling

Analysis using Monte Carlo simulation has the advantage of allowing the analysis of a system with different failure and repair distributions, thereby creating a more accurate model and better representation of system availability performance. In Monte Carlo simulation, a computer is used to generate and evaluate random variable models. In the approach presented here, a system is modeled for the duration of its life (or longer, if necessary, to obtain accurate results). This life simulation is performed for many trials to obtain a statistical result. This statistical result represents the performance of the system. Monte Carlo simulation is computationally intensive and requires significant computing power to complete all but the simplest simulations in a timely manner. Fortunately, modern computing has advanced to a point where significant computing power is readily available in off-the-shelf desktop computer platforms. The Monte Carlo simulation algorithm consists of three major steps.

1. *Simulate the State of Individual Components.* In this step, the life-cycle state of each component is modeled. The TTF and TTR are computed as random variables until the system life has been reached. This model results in a time series representing each component as a working or failed state for each sample in time.

2. *Evaluate the System State from Individual Component States.* A logic function is developed and applied to the system components to determine the operational state of the system for each time series sample. An output time series is produced representing the system state for each time series sample.

3. *Compute the Desired System Metrics (Availability, Reliability, MTTF, MTTR, etc.) from Output System States.* Performance metrics for the system are calculated using the system state time series. Metrics such as availability, reliability, MTTF, and MTTR are easily computed from the system state time series.

The simulation algorithm uses the output of each step as an input to the next step to help modularize the process. Figure 1.20 shows an overview of the algorithm process for Monte Carlo simulation.

1.7.2   Individual Component Models

Modeling of a system requires the simulation of each individual system component. Each relevant system component must be represented in order to provide an accurate assessment of system performance. Recall that relevant components are defined as those components that impact system performance when a failure occurs. Irrelevant components are those whose state does not impact the performance of the system.

1.7.2.1   Step 1. Component Description. Individual components are modeled by representing the component as a combination of two discrete random variables.

Random variable TTF simulates the time to failure of the component. Although this variable can take on any random process distribution, the failure of electronic components is typically modeled as an exponential random process. The exponential distribution is completely defined by the parameter value λ. The value λ represents the

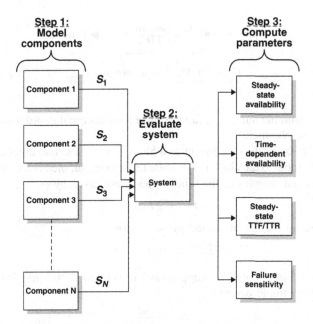

Figure 1.20. Monte Carlo system analysis algorithm.

component failure rate and is usually provided by the manufacturer or vendor of the equipment. This parameter can be specified in units of failures per hour (failure rate), failures per 10^9 h (FITs), or hours per failure (MTBF). The TTF random variable is thus expressed mathematically as

$$TTF \sim EXP(\lambda) \tag{1.65}$$

Random variable TTR simulates the time to repair of the component that follows the failure of that component. This variable can also take on any random process distribution. In system analysis, the proper selection of the repair distribution and its parameters is crucial to obtaining an accurate simulation. Many different techniques for simulating equipment repair exist. These techniques are discussed in Section 1.6.

The failure and repair random processes are sampled to produce a time domain representation of the system state based on the sampled values of TTF and TTR. The algorithm for translating these sampled values into a time domain state vector is presented in the next section.

1.7.3 Time Series Creation

Component life is modeled over a time period defined by the simulation duration requirement and is represented by the variable t_{end}. The simulation duration must be long enough to accurately assess the component availability. Highly available components can require a simulation duration longer than the system life to produce reliable

Figure 1.21. Component model.

statistics. A measurable number of failures must have occurred in order for an accurate availability to be obtained.

Figure 1.21 shows the component as a block that accepts one input and produces one output. The input to the block is a discrete time series sampled at the interval t_{sample}.

The individual samples of the times series $\mathbb{T}$ are given by

$$t_i = i \times t_{sample} \tag{1.66}$$

These samples are placed into a time series vector

$$\mathbb{T} = [\, t_0 \;\; t_1 \;\; t_2 \ldots \; t_N\,] \tag{1.67}$$

where N is the total number of samples and is defined by the required simulation duration and the sample period

$$N = \frac{t_{end}}{t_{sample}} \tag{1.68}$$

The output of the block is a component state vector $\mathbf{S}$ representing the state of the component for each time series sample

$$\mathbb{S} = [S_0 \; S_1 \; S_2 \ldots S_N] \tag{1.69}$$

The algorithm for creating the state vector output as a function of time for each component is as follows.

1. Sample TTF from failure distribution.
2. Create "system working" samples.
3. Sample TTR from repair distribution.
4. Create "system failed" samples.
5. Repeat Steps 1–4 until system life $(t \geq t_{end})$ is simulated.

The selection of t_{sample} must be such that the sampling period is sufficient to resolve all failures and repairs of the component. This sampling requirement is given by the Nyquist relation

$$t_{sample} \leq \frac{1}{2}\text{MIN(TTF, TTR)} \tag{1.70}$$

Evaluating the minimum values of TTF and TTR requires knowledge of the distribution for the random process associated with those variables. In practice, TTF $\gg$ TTR and only TTR statistics need to be analyzed. Calculation of the minimum sampled value for TTR can be performed by analyzing the distribution for TTR and selecting an appropriate sample period.

The value for t_{sample} must be selected such that

$$\text{TTR} \geq 2 \cdot t_{sample} \tag{1.71}$$

In terms of the TTR random variable PDF

$$Pr(\text{TTR} \geq 2 \cdot t_{sample}) = P \tag{1.72}$$

where P is the probability that the value of TTR will be sufficiently large to be resolved by the sampling period t_{sample}. Numerical methods can be used to determine the value of t_{sample} required to provide the desired probability P. In practice, P should be chosen such that the probability of not resolving a repair is unlikely. A poorly selected value of t_{sample} will result in a sampling error that skews the system availability to an artificially higher value. This skew is due to the unresolved repairs that do not appear as failures in the component output.

1.7.4 State Vector Creation

The state vector creation algorithm takes the time series as an input and creates a component state sample for each time series sample. The state vector defined in this procedure is a binary vector, taking on values of one and zero. The working state is given a numerical value of one and the failed state is given a numerical value of zero.

$$S(\text{working state}) \equiv 1$$
$$S(\text{failed state}) \equiv 0 \tag{1.73}$$

Figure 1.22 provides a flow chart diagram of the algorithm implementation. The details of each step are provided below.

1. *Algorithm Start.* Set the current time value to 0. This step takes as its input the time series T and the simulation duration t_{end}.
2. *Sample TTF.* Select a random value from the failure distribution model. This value represents the component time to failure.
3. *Sample TTR.* Select a random value from the repair distribution model. This value represents the component repair time. It includes fault diagnosis and repair.
4. *Current Time Iteration (TTF).* The current time t_{curr} is set to the cumulative (elapsed) time value. The cumulative time is then incremented by the failure value TTF selected in Step 2.
5. *State Vector Assignment (TTF).* The value of the state vector for all samples lying between the current time t_{curr} and the cumulative time t_{cum} is assigned the working state value $S(\text{working state}) = 1$.
6. *Current Time Iteration (TTR).* The current time t_{curr} is set to the cumulative (elapsed) time value. The cumulative time is then incremented by the repair value TTR selected in Step 3.

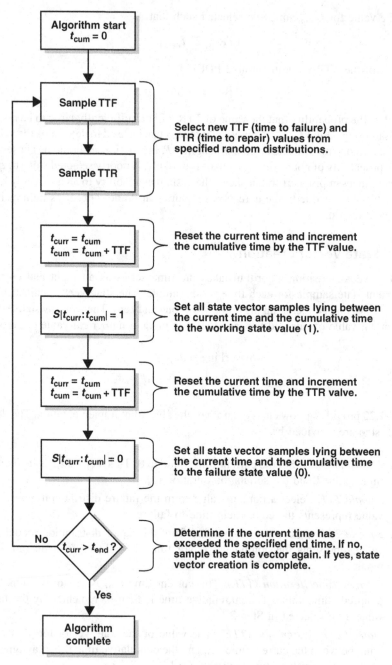

Figure 1.22. State vector algorithm flow chart.

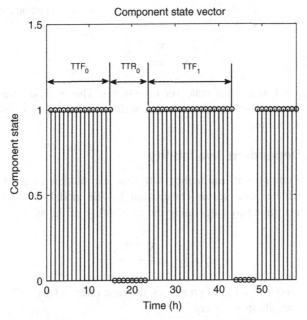

Figure 1.23. Sample state vector algorithm output.

7. *State Vector Assignment (TTR).* The value of the state vector for all samples lying between the current time t_{curr} and the cumulative time t_{cum} is assigned the failed state value $S(\text{failed state}) = 0$.

8. *Evaluate Cumulative Time.* The cumulative time value t_{cum} is compared with the end time value tend supplied to the algorithm. If the end value has not been exceeded, the procedure is repeated from Step 2. If the end time value has been exceeded, the process is complete and the algorithm ends.

A sample output (generated in MATLAB) of the state vector algorithm is shown in Figure 1.23.

1.7.5 Steady-State Availability Assessment

Steady-state availability assessment of the state vector is straightforward. The calculation can be performed directly from the output state vector. Availability is defined as the probability that a system (or in this case a component) is operating at any instant in time (see Section 1.1). The steady-state availability is the average value of the availability over the system life. This can be expressed mathematically as previously shown:

$$A = \frac{\text{item uptime}}{\text{item operational time}}$$

Numerically, the component model state vector is a binary vector in which a value of 1 represents the "component working" condition and a value of 0 represents the "component failed" condition. The availability of the component can be calculated from the state vector by applying

$$A = \frac{\sum_N^{i=0} S_i}{N} \tag{1.74}$$

where N is the total number of state vector samples. This analysis does not require knowledge of the time series vector, since availability is a ratio of "working" samples to the total number of samples.

1.7.6 Time-Dependent Availability

The variation of component availability with time can be determined from the time series vector and the state vector constructed by the model. The time-dependent availability is determined by calculating the availability of the system for each time series sample.

$$A(t_n) = \frac{\sum_n^{i=0} S_i}{n} \tag{1.75}$$

where n is the number of samples present on the time interval $[0, t_n]$. Calculating $A(t_n)$ for each sample results in the array

$$\mathbb{A} = [A(t_0) A(t_1) \ldots A(t_N)] \tag{1.76}$$

where the steady-state availability is the Nth (last) term in the array.

1.7.7 Time-to-Failure/Time-to-Repair Calculations

The TTF and TTR can be calculated for a component or a system using the time series and the state vector arrays. Calculation of these values is performed by counting the number of samples for each discrete system event. The steps involved in this algorithm are as follows.

1. Partition working and failure blocks into discrete bins.
2. Sum the number of samples in each bin.
3. Multiply the summation of those samples by t_{sample}.

Partitioning of the failures and repairs is the most difficult task in the implementation of this algorithm and will depend on programming style and the programming language chosen. Once the state sample sets have been partitioned, the TTF and TTR values are calculated by applying the equations

$$\text{TTF}_i = \sum_{n=j}^{k} (S_n = 1) \times t_{\text{sample}} \tag{1.77}$$

$$\text{TTR}_i = \sum_{n=j}^{k} (S_n = 0) \times t_{\text{sample}} \tag{1.78}$$

where j and k are the start and finish indices of the partitioned bins. The number of TTF and TTR values will vary by simulation. Averaging over the trial set is recommended to obtain an accurate assessment of the time to failure and time to repair for the components.

1.7.8   System Analysis with Multiple Components

The method for translating a set of component state vectors representing a system into a single system state vector is presented in this section. The algorithm for computing the system state vector is as follows.

1. Simulate system components using the individual component model.
2. Create a sample vector consisting of individual component sample states for time t_n.
3. Evaluate the system state for each sample state vector.
4. Calculate the availability of the system from system state vector.

This procedure assumes a general system comprised of N components. Each of the N discrete components must be modeled using the same sample period t_s so that the component state samples are correlated in time. Each component has a state vector

$$\mathbb{S}_0(t) = [S_0(t_0)\ S_0(t_1)\ S_0(t_2)\ \dots\ S_0(t_M)]$$
$$\mathbb{S}_1(t) = [S_1(t_0)\ S_1(t_1)\ S_1(t_2)\ \dots\ S_1(t_M)]$$
$$\vdots$$
$$\mathbb{S}_N(t) = [S_N(t_0)\ S_N(t_1)\ S_N(t_2)\dots S_N(t_M)]$$

(1.79)

where M is the number of samples in the system life. All values of $\mathbf{S}$ are binary (1 or 0). The system state is assessed for each sample in time. That is

$$\mathbb{S}(t_n) = [S_0(t_n)S_1(t_n)S_2(t_n)\ \dots\ S_N(t_n)]$$

(1.80)

Thus, the system state is a function of time

$$\mathbb{S}_{\text{system}}(t_n) = F(\mathbb{S}(t_n))$$

(1.81)

where $F(\mathbb{S}(t_n))$ is the rule set function and is applied to the sample set $\mathbb{S}(t_n)$.

1.7.9   System State Synthesis

Determining the state of the system based on the individual component states requires the development of a rule set that defines the state of the system for all possible sample sets. This rule set function $F()$ can be defined by developing a flow diagram relationship between the component states and the system state. In the case of simple component combinations, a mathematical relationship between the component states and the system state may be possible. Use of the flow diagram approach for more complicated systems simplifies this process as the component count grows and the interactions of the components become more complex. Definition of the rule set used for a system is highly dependent on the system and is specific to each system being modeled. As such, two simple cases are presented in the next sections. The serial combination and parallel combinations of components can be assessed using mathematical relations. More complicated systems require the flow diagram approach.

The serial and parallel cases are presented in both mathematical and flow diagram form to demonstrate the procedure.

1. *Serial Components.* The rule set for serially connected components put into words is "if any one component fails, the system has failed". Since the system working condition is defined with a numerical value of one, the state of a serially connected system is the product of the N component states. Consider a system with N individual components. The system state for these N components would be

$$\mathbb{S}_{\text{system}}(t_n) = \prod_{i=0}^{N} S_i(t_n) \qquad (1.82)$$

2. *Parallel Components.* The rule set for parallel component configurations is more complicated, since many different types of configurations of component redundancy exist. For the simple case of two components where one is required for system operation and both operate continuously, the system state rule is "if either of the components is working, the system is working." This can be expressed mathematically as the logical OR operation.

$$\mathbb{S}_{\text{system}}(t_n) = \text{OR}(S_1(t_n), S_2(t_n)) \qquad (1.83)$$

3. *Arbitrary Component Configuration.* The rule set for an arbitrary system made up of N different components can be analyzed by developing a system state flow chart that details failure flow. Although not technically required, the process for developing the system state flow chart for the serial and parallel configurations is demonstrated here for clarity. Figure 1.24 shows the system state flow chart for serially connected components.

In the case of two parallelly connected components, the flow diagram shown in Figure 1.25 is applied to determine the system state.

As can be seen in Figure 1.25, the benefit of using the flow diagram approach quickly becomes evident. Since only the outcome of the system state is of interest, the state of component 2 can be ignored if component 1 is working. This benefit is multiplied many times as the system becomes more complex. This approach implicitly applies *don't care* conditions to many component state combinations. Care must be taken such that actual failure modes are not neglected in the flow chart development.

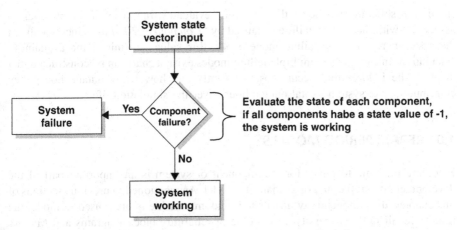

Figure 1.24. Serial component state assessment flow diagram.

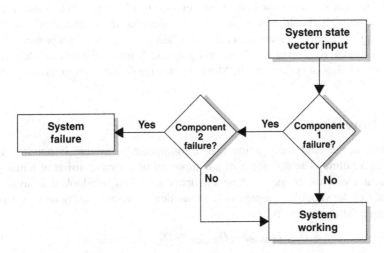

Figure 1.25. Parallel component state assessment flow diagram.

1.7.10 Failure Mode Sensitivity

When availability analysis is performed during the design phase of a project, it is desirable to know which components contribute most significantly to the unavailability of the system. A technique is presented here for quantifying that contribution. Development of the system state rule set establishes conditions on which "system working" or "system failed" decisions are made. During evaluation of these conditions, the numerical count of samples corresponding to the different failure modes can be summed. For example, in the parallel redundancy configuration presented in Figure 1.25, only one failure mode exists. This failure mode, labeled *"Component 2 Failure?"* is what causes a system failure. While this condition is being evaluated in the simulation software, whenever this condition evaluates true, a failure mode counter associated with the system failure is caused by *"Component 2 Failure"* increments. In the case where only one failure mode exists, this value will mirror the availability calculation. In cases where multiple failure modes exist, a chart can be constructed that displays the failures that occur most frequently and how many outage hours they contribute to the system's total outage hours over the simulation life.

1.8 REPAIR PERIOD MODELS

Modeling the time to repair for a component or system is an important part of the development of a reliability or availability model. As mentioned in previous sections of this chapter, the exponentially distributed random variable is often used to model the time to repair out of necessity. In the case of reliability block diagrams and Markov chain analyses, an exponentially distributed time-to-repair model is the only option due to the requirement for the memory-less condition to be met.

This section discusses methods to model time to repair and their implications on model accuracy. It should be obvious to the reader that the exponential distribution, while being a good fit for electronic component failure modeling, is not particularly well suited to model the repair of those components. When reliability models require accurate modeling of system repair, Monte Carlo simulation is often the only feasible option.

1.8.1 Downtime

Downtime is the total time period that a component or system is not functioning following a failure. The downtime of a component or system consists of a number of constituent elements. Some of these elements are often overlooked in availability analysis. Let the variable D represent the total downtime of a component or a system following a failure event. We can write D as

$$D = D_{\text{identify}} + D_{\text{dispatch}} + D_{\text{repair}} + D_{\text{close-out}} \tag{1.84}$$

When considering the downtime of a component, it is important to review and understand the service-level agreement associated with the service or system being analyzed. The total system downtime consists of at least the following elements:

D_{identify}. Downtime associated with the identification of a failure. In telecommunications systems, the time associated with identifying a failure may depend on human, electronic, or a combination of human and electronic factors. This value can be as little as seconds in the case of an electronically alarmed network operations center or could be hours for a service that requires customer feedback to identify a failure.

D_{dispatch}. Dispatch downtime is the outage time associated with travel to the location of the failure. Telecommunications systems are typically implemented with a network operation center contacting a field technician or an engineer to repair a failed component or system. The time to dispatch after the failure that has been identified can vary dramatically. In urban environments, with well-staffed technician resources, the dispatch time might be an hour or less. In remote or rural environments, the time to dispatch can be a day or more when fly outs or rural travel are required.

D_{repair}. The repair downtime is often confused with the total downtime or time to repair. Specifically, the repair downtime refers to the downtime associated with the actual repair activity. This could be the replacement of an interface module, repair of a fiber-optic cable break, or the bypass of a service to a backup configuration. The amount of time required to effect a specific type of repair can often be modeled accurately but careful consideration is in order.

$D_{\text{close-out}}$. Close-out downtime refers to the amount of time required to relay repair messages back to the appropriate parties. This downtime may be very small in systems that electronically log system up and down events. In cases where manual outage logs are contractually required, this time may have a finite and measureable effect on the total downtime. Normally, a system is returned to service immediately upon completion of repair. Examples where $D_{\text{close-out}}$ must be considered and included are, for example, the time to move traffic back to the primary system if the traffic was manually routed to alternate path. It may also be the time to achieve customer acceptance that the system is in fact repaired (the customer may want to test the repaired system and concur that it is indeed meeting performance requirements).

The four downtime elements listed above are not meant to represent a comprehensive list of all possible contributions to downtime for a system or a component. Rather, these elements are common to most repairs following the failure of a telecommunications system or a component. Each element provides a distinct contribution to the total downtime and can be modeled using a different statistical distribution (in the case of Monte Carlo model). Of particular note is the opportunity to analyze sensitivity of system performance to changes in downtime element. For example, by varying the dispatch downtime portion of the total downtime in a Monte Carlo simulation, one can glean insight into the effect of operational improvements on downtime performance.

1.8.2 Statistical Models

This chapter presented a number of different statistical models that can be used to model either the time to failure, time to repair, or both for a component or a system.

In order to better understand how to select an appropriate downtime or time-to-repair model, we will present an example.

Consider a single component where the downtime to be modeled takes on a one of four different distributions. Assume that through empirical data collection and process analysis, the following time-to-repair observations are made.

1. Mean downtime is 8 h.
2. Downtime variance is 2 h.
3. Downtime never exceeds 24 h.
4. Downtime is always greater than 1 h.

The exponential distribution is completely defined by a single parameter. The field of reliability analysis typically refers to the repair rate of an item as μ.

The PDF and CDF for an exponentially distributed time to repair with MTTR = 8 h is shown in Figure 1.26. The PDF and CDF for a normal distributed random variable with an MTTR = 8 h and variance = 2 h are shown in Figure 1.27. Recall that the MTTR is equal to $1/\mu$ for exponential random variables.

If we compare the time-to-repair models in Figures 1.26 and 1.27 to our model criteria, we find that although the mean value is a good fit, the other criteria are not a good match. Specifically, neither the target for a not to exceed value of 24 h nor the must be greater than value of 1 h are both missed. Unfortunately, in the case of the exponential distribution, one often has to modify the mean value assumptions if the not to exceed or greater than criteria are particularly important and an exponential distribution is a requirement.

The exponential distribution model for the time to repair in this example would therefore not be a particularly good fit. It may be desirable in some circumstances to proceed with the analysis but having performed the comparison shown in Table 1.5, the

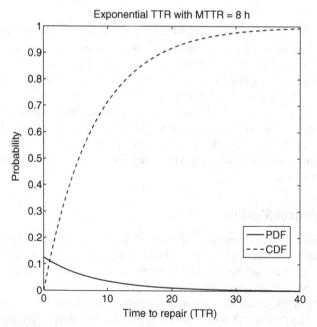

Figure 1.26. Exponentially distributed TTR with MTTR = 8 h.

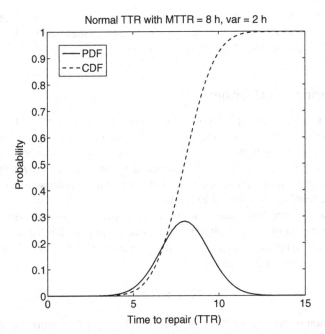

Figure 1.27. Normal distributed TTR with MTTR = 8 h, variance = 2 h.

Table 1.5. Exponential Time-to-Repair Criteria Versus Model

Criteria	Exponential Model Value	Normal Model Value
Mean downtime is 8 h	Mean downtime is 8 h	Mean downtime is 8 h
Downtime variance is 2 h	Not applicable	Variance is 2 h
Downtime never exceeds 24 h	95% of values are less than 24 h	True
Downtime is always greater than 1 h	11% of values are less than 1 h	True

limitations of this model have been clearly identified and the analyst should proceed with caution.

1.9 EQUIPMENT SPARING

The concept of equipment sparing is central to operation of telecommunications networks. In consideration of the importance of equipment sparing, it would seem obvious that careful attention should be paid to both equipment spares placement and quantities available for repair. It is unfortunate that equipment sparing design is often neglected in telecommunications systems. In many cases, sparing levels are determined by historic "experience" and are not based on quantitative analysis.

This section discusses optimization of equipment sparing levels, the impact of sparing levels, and RMA on system downtime and considerations for geographic placement of spares in long-haul systems.

1.9.1 Sparing-Level Optimization

Selection of sparing levels in telecommunications systems can be a difficult problem when quantitative analysis tools are not used. Optimization of the spares pool level for system components has important financial and logistical implications. Selecting the minimum spares pool quantity for any particular component minimizes the logistical impact of storage and management of hardware on warehouse staff while also minimizing capital or operational expenditures.

Consider a system consisting of n discrete, identical components such that $S = \{1, 2, \ldots, n\}$ is the set of all components in the system. Assume that the n components in the system S are in operation at a time $t = 0$ and operate for a duration T. Thus, the total operational time for all components is

$$T_{\text{total}} = N \times T \tag{1.87}$$

Calculation of the sparing level requires knowledge of the failure characteristics of each component. Replaceable items within the system must be identified and analyzed to determine the distribution of failures, failure rate ($z(t)$), or MTBF. Any of these three metrics can be utilized for analysis. It is most desirable to use a combination of empirically collected field data in conjunction with calculated failure rates. This provides the best combination of academic and empirical experience. Without knowledge of the failure behavior of system components, it is impossible to determine optimal sparing levels.

Assume that the MTBF of each component is given by M. The predicted average failure count for the system in time period T can thus be calculated as

$$F_{\text{sys}} = \frac{T_{\text{total}}}{M} \tag{1.88}$$

where M is the empirical or calculated mean time between failures. The average failure count can be used to determine the predicted number of spares required for each time period T. It should be noted that many telecommunications utilize maintenance agreements in which failed components are repaired by a vendor at a rate determined within a prearranged contract. This maintenance agreement can complicate sparing-level determination. The maintenance agreement contract must have specific provisions for turnaround time period in equipment repair. This turnaround time T_{vendor} must be weighed against the time between system events. The failure rate of that system can be calculated as

$$f = \frac{F_{\text{sys}}}{T} \tag{1.89}$$

where f is expressed in failures per hour. With knowledge of the failure rate f and the maintenance agreement turnaround time T_{vendor} we can calculate the required spares.

Consider a system F_{vendor} failure events per time period T_{vendor}

$$F_{vendor} = f \times T_{vendor} \tag{1.90}$$

The number of required spares N_{spares} must be greater than or equal to the number of failures expected to occur during the vendor repair period.

$$N_{spares} \geq F_{vendor} \tag{1.91}$$

In cases where a vendor maintenance agreement does not exist, the sparing levels must be selected such that a sufficient number of spares is purchased so that all failures can be repaired within a give period of time T. In this case, the calculation of N_{spares} is straightforward:

$$N_{spares} \geq F_{sys} \tag{1.92}$$

where F_{sys} is the number of predicted failures in the system over a time period T.

As an example, consider a wireless cellular network consisting of 100 base station transceiver elements. Through empirical analysis and vendor interaction, it is determined that the MTBF for the base station transceiver element is approximately 55,000 h of continuous operation. It is desired to analyze the sparing levels required for both design in which no maintenance agreement is assumed and for a system with a maintenance agreement where the turnaround time $T_{vendor} = 8$ weeks for a 1-year period (8760 h). First, we will calculate the total number of operational hours in the system consisting of 100 base station transceiver elements. The total system operational time is

$$T_{total} = N \times T = 100 \times 8760 = 876000 \text{ h}$$

The predicted average number of annual failures can thus be calculated as

$$F_{sys} = \frac{T_{total}}{M} = \frac{876000}{55000} \approx 15.9$$

Thus, the number of expected annual failures per year is approximately 16 under steady-state operation. The number of required spares for the case in which spares are annually purchased and allocated is given by

$$N_{spares} \geq F_{sys} \rightarrow N_{spares} \geq 16$$

Thus, the number of spares required for steady-state operation is 16. It should be noted that this analysis assumed average behavior. It is always good practice to select sparing levels such that anomalies can be accommodated. A reasonable sparing compliment for one year on the system above might be a value $16 \leq N_{spares} \leq 20$. Because the failure rate of the system F_{sys} is a statistical value, the number of failures in any given year can vary. The number of spares purchased in one year may be insufficient while another year it may be too great. This variation tends to disappear as the number of deployed components becomes large and the statistics become stationary in time.

In the case where a maintenance agreement exists with $T_{vendor} = 1344$ h, we must first calculate the system failure rate f as

$$f = \frac{F_{sys}}{T} = \frac{15.9}{8760} = 1.8 \times 10^{-3} \text{per h}$$

The number of failures that might occur during a vendor repair or replacement period is thus

$$F_{vendor} = f \times T_{vendor} = \left(1.8 \times 10^{-3}\right) \times 1344 \approx 2.4$$

By applying the spares count rule, we find that

$$N_{spares} \geq F_{vendor} \rightarrow N_{spares} \geq 3$$

Clearly, in the maintenance agreement model, the number of spares required is significantly smaller than in the self-repaired case. The trade-off analysis between maintenance agreement costs and the equipment costs is now easy.

Assume that the base station transceiver element has an equipment cost of $40,000 per element and that the annual maintenance agreement cost is $350,000 (Table 1.6).

The maintenance agreement approach to this particular problem is clearly the less-expensive solution. Although a telecommunications provider may opt to select a self-repaired model for finance or business reasons, it is easy to see the cost trade-offs after the sparing analysis is complete.

1.9.2 Geographic Considerations for Spares Placement

Analysis of the geographic placement of spare components is often required in order to achieve the required time to repair for systems covering large geographic areas or having very difficult terrain.

Telecommunications networks generally cover large geographic areas due to the nature of their mission. Whether the system is a long-haul submarine fiber-optic network, a backbone microwave system, or an urban cellular wireless network, the area being served typically covers a large geographic region. Because of this large area being served, it is important to consider the optimal sparing levels and placement to ensure that both the time to repair and the number of spares available maintain the necessary levels.

Table 1.6. Spares Cost Comparison Between Self-Repaired and Vendor-Repaired Models

	Self-Repaired System (No Maintenance Agreement)	Vendor-Repaired System (Maintenance Agreement)
Spares cost	16 × $40,000 = $640,000	3 × $40,000 = $120,000
Maintenance agreement cost	N/A	$350,000
Total annual cost	$640,000	$470,000

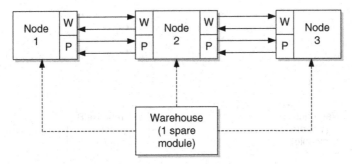

Figure 1.28. Centralized warehousing and dispatch sparing approach.

A number of different approaches for locating spare equipment are provided below:

1. *Centralized Warehousing and Dispatch.* In a centralized warehousing approach (Figure 1.28), all system spares are located in a central warehouse or depot and are picked up or shipped from this location in the event of a failure. For systems implementing full redundancy, this approach is often the most convenient since the time to repair can be relaxed enough (because of component redundancy) to support the logistics time required to place a spare unit on site. In cases where the shipping or logistics time causes the time to repair to exceed the requirement, this approach may be unacceptable.

   Systems implementing a relatively small number of deployed components can also benefit from a reduced spare equipment count. Consider a fiber-optic network consisting of a total of eight optical interface modules (four working, four protection). Assume that the sparing level analysis results in a requirement for one spare optical interface module. In the centralized sparing model, only one spare interface module would be purchased and placed in the warehouse. This module would be deployed when any failure occurs in the system.

2. *Territorial Warehousing and Dispatch.* Territorial warehousing places spare equipment at strategically selected locations, reducing the logistics time to place units on site while keeping the spare unit costs at a reasonable level. In the case of a unit failure in the system, the spare unit would be dispatched from a predetermined location that provides the minimal logistics dispatch time.

   Examination of the system presented above using a territorial approach to sparing results in an increased spare unit requirement of one additional spare is shown in Figure 1.29. Warehouses A and B would both store one spare optical interface module each. If Node 1 was at a significant geographic distance from Nodes 2 and 3 (e.g., in a submarine fiber-optic network), this approach would represent a good compromise of performance versus cost.

3. *On-Site Sparing.* The last sparing approach to be considered is the on-site sparing model (Figure 1.30). In this model, every site houses the spares required to restore the system in the case of an outage or failure. This approach to sparing

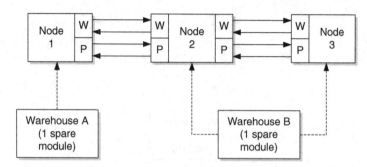

Figure 1.29. Territorial warehousing and dispatch sparing approach.

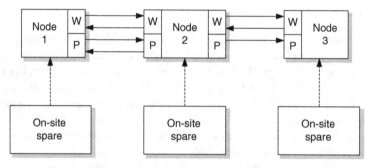

Figure 1.30. On-site sparing approach.

provides the highest attainable performance since the logistics time to place a spare on site is zero. On-site sparing comes at the highest cost as well. It is common in systems operating with on-site spares to see a dramatically increased sparing cost because of spares redundancy required to place spare equipment on site. In the case of the example presented here, on-site sparing would result in three spare interface modules (one at each location). This is three times the sparing level calculated due to expected failures.

QUESTIONS

1.1. Create a flow chart that graphically depicts the development of reliability engineering in the twentieth century.

1.2. Modern reliability engineering analysis utilizes what type of mathematics for analysis? What is the purpose of this type of mathematical analysis?

1.3. What role does empirical data play in modern reliability analysis? What specific implications does reliability engineering have on telecommunications systems?

1.4. Enumerate your goals as a reader of this book. What benefits do you hope to derive from the study of telecommunications reliability analysis?

1.5. Define the term *reliability* and give three practical examples of reliability applications. Ensure that the examples provide both the duration and the conditions for which the reliability is defined.

1.6. Describe the time to failure and its relationship with the reliability function.

1.7. Define *availability* and give three examples of its use in telecommunications design, operation, and business.

1.8. Explain the difference between average and instantaneous availability. Provide examples where the average and instantaneous availability are the same and are different.

1.9. A particular network element's datasheet indicates an MTBF of 65,000 h. Calculate the network element availability (in percent) if the expected mean downtime is 12 h.

1.10. If 25 of the network elements described in Q1.9 are placed into service at time $t = 0$, what is the expected number of element failures annually?

1.11. Explain *maintainability* in terms of an operational telecommunications system. Why is the maintainability metric a critical performance measure? Describe qualitatively how downtime and maintainability are related.

1.12. A vendor provides an MTBF in their equipment cutsheet indicating a value of 125,000 h. Convert the MTBF to both failure rate (in failures/h) and FITS.

1.13. A system of 100 telecommunications nodes is deployed and operates for 5 years. The table below enumerates the annual failures per year. Calculate the annual failure rate (in failures/h and FITS) for each year and the average failure rate (in failures/h and FITS) for the 5-year period.

Year	Failures
1	6
2	4
3	8
4	7
5	3

1.14. An interface card for a multiplexer has an MTBF of 95,000 h as defined on a vendor datasheet. Assuming that the TTF for the card is exponentially distributed, write the TTF, PDF, and CDF functions. Plot the PDF and CDF functions using a graphing calculator or computer analysis tool.

1.15. Why is an exponentially distributed random variable beneficial for analyzing telecommunications systems hardware? What is the failure rate of an exponentially distributed random variable.

1.16. Why is the exponential distribution poorly suited for modeling time to repair? What distributions are well suited to model system downtime?

1.17. The time to repair of a network is characterized by a mean value of 24 h. Assuming that 90% of the variability of the TTR is contained with the range of 12–36 h, develop normal, lognormal, and Weibull distribution models for the TTR. Plot the PDF of each distribution.

1.18. Empirical data collection has tabulated the date and downtime for repair of a system. Develop a Weibull TTR model for the data shown in the table below. Plot the CDF and PDF of the model developed. Calculate the MTBF of the system.

Date	Downtime (h)
6/15/2001	11
9/3/2002	3
12/5/2002	14
7/5/2003	5
11/2/2003	9
2/6/2004	20
4/29/2004	2
8/4/2004	7
10/21/2004	3
12/16/2004	13

1.19. What flexibility does the use of Monte Carlo simulation provide in system analysis? What are the advantages of using reliability block diagrams or Markov chains?

1.20. Define "relevance" as it relates to a reliability block diagram analysis. What is the impact of a relevant component on the reliability performance of a system? What impact does an irrelevant component have on system performance?

1.21. A telecommunications network consists of three discrete components that are combined to form a single-thread network. If the reliability of each constituent component is 99.9%, 99.99%, and 99.95%, respectively, what is the maximum achievable reliability of the system (based only on observation of the constituent component reliabilities and without performance a calculation)?

1.22. Applying the serial combination structure function definition for system reliability, calculate the actual reliability of the single-thread system described in Q1.21.

1.23. Calculate the reliability of the following two system designs.

a. Serial combination of 1:1 redundant components in Q1.21.

b. Parallel combination of serial components in Q1.21.

1.24. Redundancy is being considered for a telecommunications subsystem. If the modular system costs $15,000 per module and four active modules are required, calculate the following (assume that each module has an MTBF of 40,000 h).

a. The cost difference between a 1:1 and a 1:4 system design.

b. The reliability after 2 years for each system.

1.25. Explain why reliability is not applicable as a performance metric in repairable system analysis. For what types of systems is reliability a good metric?

1.26. Describe Monte Carlo simulation and give a specific example of a simulation.

1.27. Describe the Markov chain analysis technique in your own words. What condition must exist in order for a process to possess the "Markov property." For what conditions is the Markov chain analysis technique best suited within a telecommunications environment?

1.28. Develop a Markov transition diagram for a system consisting of two redundant components operating in a hot-standby configuration with the same failure rate. Assume that both components have a failure rate of $\lambda = 4 \times 10^{-5}$ failures/h. Repair of each component takes 16 h on average. Indicate the failure rate and repair rate of each transition. Assume that a repair of a system failure returns the system to fully redundant operation.

1.29. Write the transition rate matrix for the transition diagram developed in Q1.28.

1.30. Assuming a steady-state solution, solve the Chapman-Kolmogorov equations to determine the probability of state occupation for the states identified in Q1.28.

1.31. Determine the availability and unavailability of the system described in Q1.28 using the results of Q1.30.

1.32. Develop a Monte Carlo simulation for the system identified in Q1.28. Assume a system life of 8 y for the simulation. Model the repair as a random variable TTR $\sim$ NORM(16, 2). Provide an analysis algorithm overview indicating the system components, evaluation logic, and metrics to be computed.

1.33. Develop a system state flow diagram for the operation of the system model in Q1.32. Implement logic to compute the state of the system for the two input system.

1.34. Simulate the system in Q1.33 for 5000 sample life cycles. Compute the life-cycle availability. Provide a histogram plot of availability.

1.35. Compare the Monte Carlo and Markov chain results. What are the simulation differences and similarities?

2

FIBER-OPTIC NETWORKS

Modern fiber-optic networks achieve throughputs and distances once thought impossible. It is now commonplace to find terrestrial and submarine fiber-optic networks with multiple 10 Gbps wavelength-division multiplexed links. Exhaustion of capacity growth within the constraints of current on–off signaling technology has brought about the advent of multiphase modulation over fiber-optic cables. This multiphase modulation allows further increase in channel capacity and total throughput.

Chapter 2 is divided into two sections. These sections address the distinct reli-ability analysis challenges presented by both the network topologies and the communications channels of terrestrial and submarine networks.. Each section is organized so that the relevant topology and common network designs are presented first, followed by the preferred reliability analysis technique and any notes or comments relevant to analysis of that system type.

2.1 TERRESTRIAL FIBER-OPTIC NETWORKS

Terrestrial fiber-optic networks encompass a wide variety of different network sizes and topologies. These network types include:

Telecommunications System Reliability Engineering, Theory, and Practice, Mark L. Ayers.
© 2012 by the Institute of Electrical and Electronics Engineers, Inc. Published 2012 by John Wiley & Sons, Inc.

- metropolitan area networks
- campus area networks
- long-haul networks

All of these network types typically utilize the same network technologies for multiplexing and transmission.

The analysis of terrestrial fiber-optic network availability often requires a detailed understanding of not only the individual node availability but also the network topology.

2.1.1 Terrestrial Fiber Paths

The fiber-optic paths interconnecting terrestrial networks utilize one of two physical installation methods. Terrestrial fiber-optic cables utilize either aerial or ground (buried) installation techniques. Aerial cables are placed on a series of supporting poles and represent a very cost-effective and rapid method for deployment. Aerial fiber-optic cable installations are subjected to a number of statistically random failure events. Acts of vandalism, pole damage, and construction equipment damage all regularly cause aerial fiber-optic cable outages.

Buried cable installation techniques result in different reliability performance than the aerial case. Shallow-buried installation techniques result in rapid deployment at low cost but are most exposed to cable damage from outside influences. Unless the cable location is remote, shallow-buried cable installation can result in frequent outages due to cable damage. Figure 2.1 shows a photograph of a shallow-buried laid (commonly called shallow trench) fiber-optic cable in western Alaska.

Figure 2.1. Shallow-buried fiber-optic cable installation example in western Alaska.

Because of the remote nature of the cable deployment, the typical cable damage concerns did not apply in this particular case. It is very common for fiber-optic cables to be buried beneath the surface of the ground in populated areas. Common causes of outages in both deep- and shallow-buried terrestrial cables include construction damage due to digging (often referred to as "backhoe fade"), river washouts, and roadway or rail accidents where the fiber is buried within the right of way of the roadway or railway. In cases where fiber-optic cabling is exposed, the fiber can also be subject to damage cause by wildlife.

Reliability and availability modeling of terrestrial fiber-optic cable paths can be challenging. As will be shown in this section, the reliability and availability of most fiber-optic networks is frequently reliant on the performance of the fiber-optic paths. Because the fiber-optic path failure performance often drives the overall system performance, it is particularly important that the time-to-failure (TTF) model, adopted for fiber-optic paths, is both accurate and reasonable. The unperturbed failure rate of the fiber-optic cable (including conduit, buffer tubes, sheath, and glass fibers) itself is extremely low. Because fiber-optic cables are factory tested and are passive network elements, they are inherently very reliable. Failures involving cable splices and patch panels are the only practical instances of fiber cable failures in real systems. Installation quality control issues can lead to fiber failures due to microbending or ice crushing where water has been allowed to enter the conduit. Cable outages due to external damage are far more frequent and should be modeled using a statistical time-to-failure model based on empirical data. The most accurate empirical data are collected from actual operational experience. In cases where operational empirical data are not available, an effort should be made to collect empirical data for a system similar to the one being analyzed. Assumptions are often made regarding the frequency of cable cut events and the time to repair (TTR) these failures. If these assumptions are not accurate, the resultant system performance predictions will be incorrect due to the sensitivity of fiber-optic network reliability performance to path model assumptions.

Table 2.1 shows an example of terrestrial fiber-optic cable event empirical data and the relative frequency of those events.

Figure 2.2 shows the empirical distribution of cable outage time to repair and time to failure for the events in Table 2.1. For the modeled TTF distribution, assume that the

Table 2.1. Sample Empirical Tabulation of Terrestrial Fiber-Optic Cable Events

Number	Date	Description	TTF (h)	Outage Duration (h)
1	3/1/2005	Cable cut due to lack of locate	1416	12.2
2	5/12/2005	Rifle shot aerial cable damage	1728	26.9
3	6/22/2005	Cable cut due to lack of locate	984	7.8
4	9/3/2005	Cable cut due to backhoe digging	1752	9.3
5	11/6/2005	Cable cut due to train derailment	1536	18.4

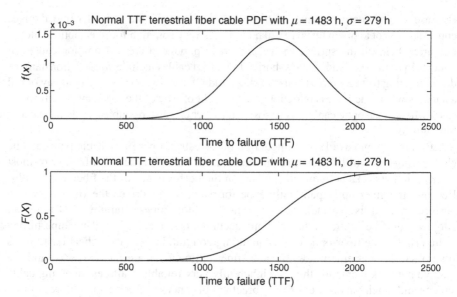

Figure 2.2. Terrestrial fiber-optic cable TTF model PDF and CDF.

TTF is a normal distributed random variable. We can calculate the mean and standard deviation for the TTF as:

$$E[\text{TTF}] = 1483\,\text{h}, \sigma = 279\,\text{h}$$

The mean and standard deviation of the TTR are easily calculated from the outage duration statistics

$$E[\text{TTR}] = 14.5\,\text{h}, \sigma = 7.4\,\text{h}$$

Unfortunately, due to the high variability of the cable repair time, we need to use a more carefully selected TTR repair model. If we apply some constraints on the model, we can develop a reasonable representation of the terrestrial cable TTR. Assume that no cable repair will occur in less than 2 h (TTR > 2 h for all TTR_n). Also assume that all fiber cable repairs will be completed in 48 h (TTR < 48 h for all TTR_n). We can select a Weibull distribution using a statistical distribution fit that reasonably approximates this behavior. Figure 2.3 shows a sample PDF and cumulative distribution function (CDF) used to model the time to repair the fiber-optic cable path. This model was selected by using a curve-fit algorithm to select the Weibull distribution scale and shape parameters.

The MTBF and MTTR of the sample fiber-optic cable path can be calculated by determining the expected value (or mean) of each distribution. Although the mean value of the cable path availability is interesting, it is often more insightful to analyze cable models using Monte Carlo simulation techniques. The results of a sample analysis using the TTF and TTR models above are shown in Figure 2.4.

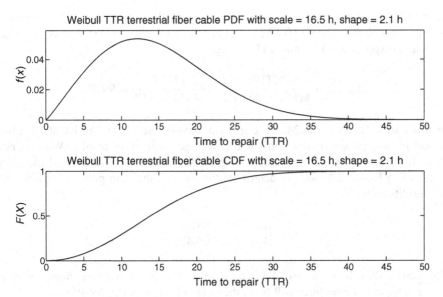

Figure 2.3. Terrestrial fiber-optic cable TTR model PDF and CDF.

The advantage of the Monte Carlo method is that it produces a distribution of observed availabilities as an output instead of a single average availability value. The analyst can thus calculate other availability performance metrics such as percentile, median, and mode quickly and easily for a relatively complex system model using

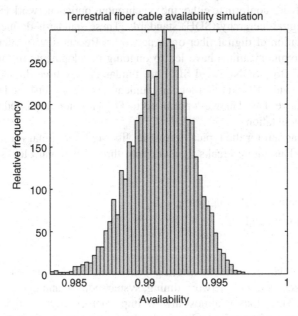

Figure 2.4. Monte Carlo simulation results for terrestrial fiber-optic cable.

Monte Carlo methods. In this case, the mean availability can be determined from the population statistics to be 99.1%. We calculated the availability using the more common approach (exponential TTF and TTR) using

$$A_{\text{average}} = \frac{\text{MTBF}}{\text{MTBF} + \text{MTTR}} = \frac{1483.0}{1483.0 + 14.5} = 99.0\%$$

In this particular case, the Monte Carlo simulation results are consistent with the closed-form exponential solution for the average availability results. What is not clear from the analysis is the range of possible values. By examining histogram statistics in Figure 2.4, we can determine the 5th and the 95th percentile values of availability to be

$$A_{5\text{th}} = 98.7\%$$
$$A_{95\text{th}} = 99.4\%$$

This bounds the achievable availability for terrestrial fiber-optic cabling. Ninety percent of the achieved availabilities will be in the range of $98.7\% \leq A \leq 99.4\%$.

2.1.2 Terrestrial Fiber-Optic Equipment

Terrestrial fiber-optic networks make up the backbone of modern high-speed communications systems. All fiber-optic networks consist of two or more nodes that are interconnected by fiber-optic paths (as described above). The majority of fiber-optic networks worldwide conform to either the synchronous optical network (SONET) or the synchronous digital hierarchy (SDH) standards. These standards define the operation and implementation of digital fiber-optic networks. Recent developments in Internet protocol (IP) communications have led to ongoing developments in the SONET and SDH standards. Both SONET and SDH are fundamentally time-domain multiplexed (TDM) systems while IP and Ethernet communications are packet-based protocols. The adaptation of packet-based network protocols to TDM topologies has led to a number of different implementations.

A block diagram for the typical terrestrial fiber-optic terminal (or node) is shown in Figure 2.5 . This figure breaks the node into three functional blocks. These blocks are:

- system control and monitoring,
- line interface, and
- drop interface.

The system control and monitoring functional block consists of the system components associated with network timing, system control and alarming, and system administration. This block is almost always implemented with redundant modules in carrier-class fiber-optic terminals. Enterprise-level terminals often forego redundancy

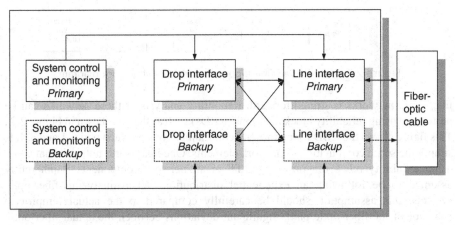

Figure 2.5. Terrestrial fiber-optic terminal functional block diagram.

for system control and monitoring to save cost. The line interface block consists of the optical interfaces associated with transmission and reception of signals through the fiber-optic cable communications channel. Common line rates include OC-192, OC-48, OC-12, and OC-3. This functional block also includes any equipment associated with wavelength-division multiplexing (WDM). The drop interface block consists of customer- and network-facing interfaces that are multiplexed by the terminal to the line rate. These interfaces often range from OC-48, OC-12, OC-3, DS-3, and DS-1 to Ethernet and others.

Many different fiber-optic network topologies exist including unprotected and protected designs.

2.1.2.1 Unprotected Fiber-Optic Networks.
Unprotected designs employ a single fiber-optic path between two or more nodes. Figure 2.6 shows an example of a three-node unprotected fiber-optic network.

In an unprotected fiber-optic network, a failure of any line interface (transceiver) or path results in a network outage. For example, in Figure 2.6, a failure of fiber paths 1 or 2, or the failure of any of the transceivers will result in a complete communications failure between nodes 1 and 3. Transceivers in fiber-optic networks are often given the designation of East or West to indicate logical directionality. East-facing line interfaces communicate logically with West-facing interfaces. The analysis of unprotected

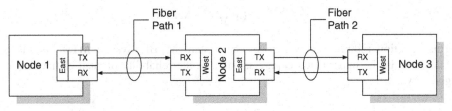

Figure 2.6. Unprotected fiber-optic network system block diagram.

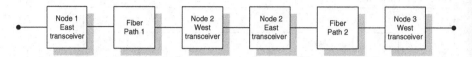

Figure 2.7. Unprotected fiber-optic network reliability block diagram.

fiber-optic networks is straightforward. Reliability analysis of these networks can be performed using reliability block diagrams (RBDs) such as the one shown in Figure 2.7. This figure shows the RBD for end-to-end communications between nodes 1 and 3. It is simply the serial combination of network elements. One item worth noting is that when using reliability block diagrams, the TTF of all network components is inherently assumed to be following an exponential distribution. When analyzing fiber-optic systems, this assumption should be carefully compared to the actual (empirical) behavior of the fiber-optic path. Significant deviations between the actual fiber-optic path TTF and the exponential distribution model can result in incorrect results. Assuming that the exponential distribution reasonably represents the cable failure TTF, we can calculate the reliability of the system as follows.

The overall network reliability can be calculated using the serial component reliability equation:

$$R(S(t)) = \prod_{i=1}^{n} R_i(t)$$

where $R_i(t)$ is the reliability of the ith system element. Thus, $R_{\text{system}}(t)$ is given by

$$R_{\text{system}} = R_{\text{Node1\_TxRx}} \times R_{\text{Fiber Path 1}} \times R_{\text{Node 2}_{TxRx}} \times R_{\text{Node 2}_{TxRx}} \times R_{\text{Fiber Path 2}} \times R_{\text{Node 3\_TxRx}}$$

If we assume that all of the optical interfaces modules are identical, the reliability simplifies to

$$R_{\text{system}} = R_{TxRx}{}^4 \times R_{\text{Fiber Path 1}} \times R_{\text{Fiber Path 2}}$$

The calculation of system availability is also straightforward in the unprotected case. In order to calculate the system availability, we will apply the following equation for average availability

$$A_{\text{Average}} = \frac{\text{MTBF}}{(\text{MTBF} + \text{MTTR})}$$

This equation requires knowledge of the system MTBF. System MTBF is easily calculated in the unprotected case. Assume that the reliability of each component is given by

$$R_n(t) = e^{-\lambda_n t}$$

The system reliability is

$$R_{sys}(t) = \prod_{i=1}^{n} R_i(t) = e^{-\lambda_1 t} \times e^{-\lambda_2 t*} \ldots e^{-\lambda_n t} = e^{-t(\lambda_1 + \lambda_2 + \ldots \lambda_n)}$$

Thus, the system failure rate (assuming exponentially distributed component time to failures) is

$$z_{sys} = \lambda_1 + \lambda_2 + \ldots \lambda_n = \sum_{i=1}^{n} \lambda_i$$

and the system MTBF is

$$\text{MTBF}_{sys} = \frac{1}{z_{sys}}$$

We can calculate the average system availability to be

$$A_{sys} = \frac{\text{MTBF}_{sys}}{(\text{MTBF}_{sys} + \text{MTTR})}$$

Note that the solution for A_{sys} shown above assumes that the MTTR for each component is identical. From a practical standpoint, this is not likely to be the case. Accurate treatment of this analysis would include separate treatment of the hardware and the fiber cable elements.

2.1.2.2 Protected Fiber-Optic Networks. Protected fiber-optic networks typically employ one of two core approaches. The network uses either a variant of SONET/SDH ring network protection (UPSR, BLSR, etc.) or a mesh networking technology. Interface directionality is important in ring-protected networks such as unidirectional path switched ring (UPSR) and bidirectional line switched ring (BLSR). In these systems, two counter-rotating rings are constructed at either the *path* or the *line* level (path and line are SONET standard references to service endpoints between nodes. For a complete understanding of path and line in the context of fiber-optic networks, the reader is encouraged to study the SONET standards). Figure 2.8 shows the protected UPSR topology in its normal operational state. The bold lines in Figure 2.8 indicate the *working* path while the lighter lines indicate the *protect* path of the ring protection scheme. Redundant data are sent along both paths at all times so that a failure of any path or transceiver can be accommodated within 50 ms of the event occurrence.

In a UPSR network, two types of link failures can occur. Each failure results in a distinct protection action being taken by the network. In the case of a fiber path failure, UPSR logic causes a system "foldback" in which traffic is routed around the failed path allowing all communications between nodes to continue to flow. There are two major implications to this logic. It requires sufficient capacity to be available on both counter-rotating rings so that in case of a failure, capacity exists to restore the traffic. Some

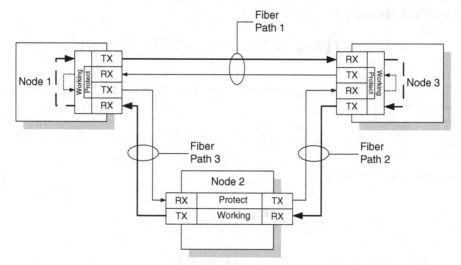

Figure 2.8. UPSR ring network topology, normal operation.

operators use the capacity of the counter-rotating ring under normal operations to double the capacity of the ring, but must forfeit that capacity upon a primary ring interruption. UPSR protection also increases the latency of any service during a foldback protection event. This latency difference can be relatively small in metro-politan or in campus networks but can be large in long-haul terrestrial networks. Latency impacts should be considered when using any fiber-optic network protection scheme so that service-level agreement contracts are not violated during a protection event. Figure 2.9 shows the UPSR ring in the condition where a fiber-optic path has failed. Note that the foldback data path is indicated in the figure by dotted lines.

Failure of a transceiver (or less commonly a single fiber) will result in a protection switch of the system from the *working* path to the *protect* path as indicated in Fig-ure 2.10. In this case, the service latency of the system is not affected and the system switching event is less than 50 ms. The bold line in Figure 2.10 indicates that the *protect* path carries traffic during this event. For maximum availability performance, the individual fiber-optic links on each fiber path should be implemented using physically diverse fiber routing (specifically, the East and the West routes should not be coincident).

Reliability and availability analysis of ring-protected fiber-optic networks can be challenging. The difficulty in analysis stems from the number of failure modes (which grows exponentially as the number of nodes increases). Monte Carlo methods are best suited to availability analysis of protected fiber-optic networks because Markov chain analyses limit the accuracy of the results (because fiber paths must be analyzed using exponential time to failure distributions) and are typically just as difficult to apply as the Monte Carlo method. Additionally, all repairs in the Markov chain model are exponentially distributed while the Monte Carlo simulation allows the repairs to be modeled by the distribution that best fits the repair of that component.

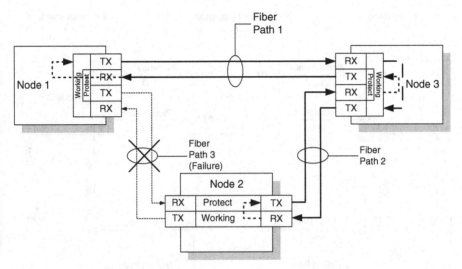

Figure 2.9. UPSR ring network topology, fiber path failure.

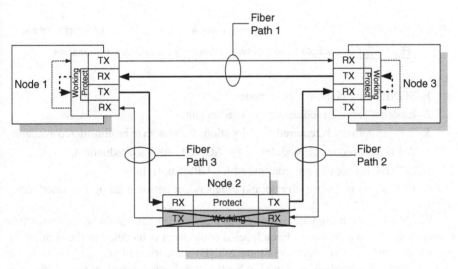

Figure 2.10. UPSR ring network topology, transceiver failure.

2.1.3 SONET Fiber-Optic Network Analysis Example

Consider a SONET network with the system topology in Figure 2.11.

The system in Figure 2.11 consists of six identical SONET nodes and utilizes a UPSR ring protection scheme. The following assumptions are made for the purposes of simulation. These assumptions would be determined through system requirements specification in a real network simulation.

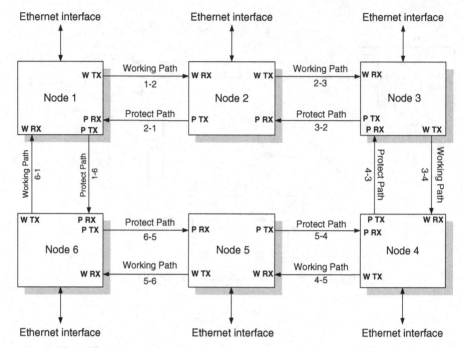

Figure 2.11. Example SONET network topology for Monte Carlo analysis.

1. SONET ring consists of six nodes.
2. Ethernet service modules are not redundant.
3. Service delivery is required to all locations for system to be considered available.
4. All common control modules in the SONET shelf are redundant.
5. SONET line optics are configured for UPSR operation.
6. Fiber-optic paths are diverse and independent between all optical interfaces.

This system is representative of a typical SONET ring that might be used to deliver Ethernet service to a customer. Each labeled component is modeled in the simulation. The Monte Carlo simulation model for the SONET network in Figure 2.11 is developed by following the procedure presented in Section 1.3. Each component is simulated for the duration of its simulation life. Determination of the components to be modeled is an important step in model development. Inclusion or exclusion of components defines both the complexity and the accuracy of the system model. The system state is then determined using the system state rule set algorithm. The discrete system components are listed below in Table 2.2.

The values in Table 2.2 are not representative of any particular equipment manufacturer but, rather, are typical values for a SONET node and its associated peripherals. All electronic equipment is modeled using an exponential distribution with a constant failure rate. The shelf value includes all equipment common to the shelf and

Table 2.2. Sample SONET Network Model Components

Component	Instances	MTBF (h)	MTTR (h)	σ (h)	Failure Distribution
SONET shelf	6	800,000	6	2	Exponential
SONET line optic	12	250,000	6	2	Exponential
Ethernet module	6	250,000	6	2	Exponential
Fiber-optic path	12	25,000	12	4	Weibull

assumes redundancy on common equipment. Fiber-optic paths are assumed to be diverse and independent with a mean time between failures of approximately 3 years. Fiber-optic paths are modeled using a Weibull distribution with a shape parameter equal to 3 and a scale parameter equal to 25,000 h. All repair distributions are assumed to be normal distributed with mean and standard deviation as shown in Table 2.2.

2.1.3.1 Simulation Parameters. In order to simulate the system accurately, the simulation duration, simulation sample period, and number of trials must be selected. Selection of the simulation duration parameter requires a value that will provide the desired result. If the intent is to determine the steady-state availability, then the duration must be long enough to observe a measurable number of failures. This duration may be much greater than the actual expected system life. This value is determined empirically by analyzing simulation results and increasing or decreasing the duration required. Simulations resulting in availabilities equal to one are assumed to not have achieved steady state since an availability of one is not achievable in a real system with real components. The duration is extended until the statistics are well behaved. The empirically determined value for simulation duration of the example system is 262,800 h (30 years). The maximum value of simulation sample period is determined by (see Section 1.3):

$$t_{sample} \leq \frac{1}{2} \text{MIN}(\text{TTF}, \text{TTR})$$

In practice, it is desirable to select a sample period somewhat less than the maximum in order to provide greater resolution. The sample period for the example system is

$$t_{sample} = \frac{\text{NORMINV}(0.05, 6, 2)}{10} = 0.2710 \, \text{h}$$

where NORMINV is the inverse normal function, 0.05 is the probability of the sample period being exceeded (as defined in Section 1.3), 6 is the minimum MTTR value, and 2 is the minimum MTTR standard deviation. The sample period is divided by a factor of 10 to provide improved simulation resolution. The number of trials used in the simulation is also determined empirically. This value must be selected such that the resultant statistics are a complete representation of system performance. Empirically, simulation results show that a value of approximately 10,000 trials is sufficient to determine steady-state availability for this system. Thus, a single simulation will consist of system state samples evaluated at 0.271-h increments for a duration of 30 years. This simulation is performed 10,000 times and the population statistics of the system's performance are computed.

2.1.3.2 System Model Rule Set. Each component is simulated using the defined parameters to generate a component state vector. Each of the state vectors is provided as an input to the system model rule set to determine the system state for each simulation sample. Development of the system model rule set is done using a flow chart approach. This flow chart is then implemented in software using program flow control statements. Figure 2.12 shows the flow chart for the SONET UPSR network. Changes to the protection, topology, or path independence assumptions would require a new rule set to be developed.

2.1.3.3 Simulation Results. Results from the simulation of the network availability are provided in Figure 2.13 . Two charts are used to show the results of the simulation. The first chart shows a histogram of the distribution of availability. It can be seen from the histogram that the availability of the system varies based on how many failures the system encounters during its lifetime.

In the second chart, the failure sensitivity of the system is shown. This sensitivity indicates that the redundant path and optical equipment are not the limiting factor in the availability of the system, but rather the Ethernet modules are the component that most greatly contributes to the system unavailability. Improvements to the system availability should be focused on adding redundancy or improving the reliability of the Ethernet module to increase system performance.

Analysis of other ring network topologies (such as BLSR or mesh) would follow a very similar approach.

2.2 SUBMARINE FIBER-OPTIC NETWORKS

Submarine fiber-optic networks constitute a specialized and important field of tele-communications engineering. The submarine networks are constructed using applica-tion-specific components such as repeaters, power feed equipment (PFE), and line terminating equipment (LTE). Long-haul terrestrial fiber-optic networks often employ a subset of the components present in the submarine networks. Due to the high-availability performance requirements, the high cost of repairs and the significant TTR associated with component failures in submarine systems reliability are often given great importance during the design and construction of submarine fiber-optic networks. The block diagram shown in Figure 2.14 shows the major components of a modern submarine fiber-optic network.

The major components making up a submarine fiber-optic network consist of:

- line terminating equipment
- tributary interface equipment
- power feed equipment
- fiber-optic cabling
- submarine repeater equipment

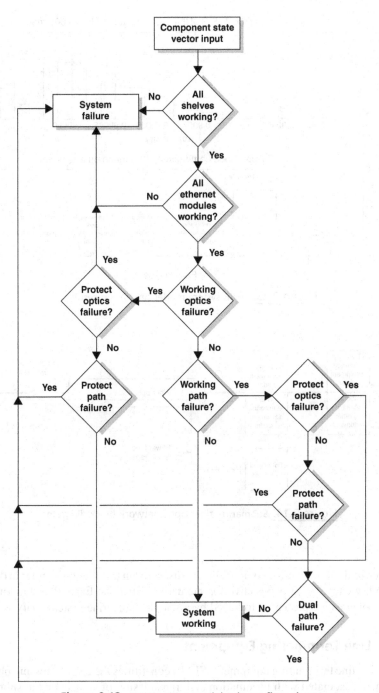

Figure 2.12. UPSR system model rule set flow chart.

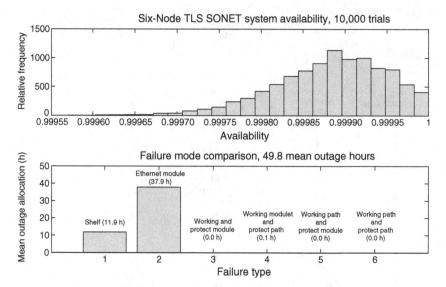

Figure 2.13. UPSR system model simulation results.

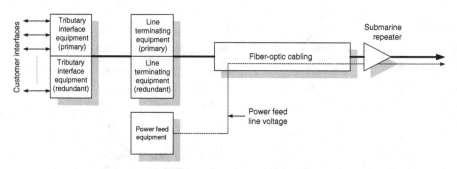

Figure 2.14. Submarine fiber-optic network block diagram.

Each of these submarine components are discussed in this section in order to develop a reliability engineering perspective for submarine networks. Example submarine network topologies are presented followed by a sample submarine system analysis.

2.2.1 Line Terminating Equipment

Submarine line terminating equipment (SLTE) constitutes the electronics and physical equipment associated with termination and transmission of signals on the submarine fiber-optic cable. Figure 2.15 shows a block diagram representative of typical line terminating equipment. Line terminal equipment includes wavelength-division

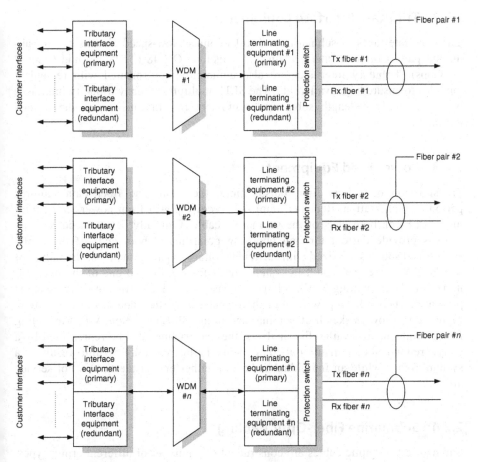

Figure 2.15. Submarine line terminal equipment functional block diagram.

multiplexing, optical amplifier (often erbium-doped fiber amplifier or EDFA), dispersion compensation, and line coding equipment. Tributary interfaces (OC-n) are multiplexed together using the line terminal equipment onto a range of wavelengths defined by the WDM equipment.

Reliability modeling of the line terminating equipment can be performed using electronic system models. All components within the SLTE are implemented in a combination of active and passive semiconductor devices. Vendors typically provide reliability performance data prior to purchase of submarine systems but often do not disclose performance without nondisclosure agreements. Almost all active line interface subsystems are implemented using redundancy to ensure system availability. The resultant SLTE equipment has very high availability and is rarely the cause of service-affecting outages.

2.2.2 Tributary Interface Equipment

Tributary interfaces in submarine networks connect low-speed circuits to the sub-marine path. Common interface line rates include OC-48 (2.5 Gbps) and OC-192 (10 Gbps). Tributary interfaces are also commonly implemented with redundant modules to ensure high availability. The SLTE multiplexes many tributary interfaces onto the multiwavelength submarine equipment for transmission to the far-end terminal.

2.2.3 Power Feed Equipment

Transmission of optical light pulses along the submarine fiber path requires periodic amplification. Repeater amplifiers receive power from power feed equip-ment (PFE) on both ends of the submarine cable. Carefully balanced power feed sources provide current through opposing polarities at both ends of the cable. Power feed sources are sized with sufficient voltage capacity such that the failure of one PFE unit does not result in system power failure. Should one of the two PFE units fail, the remaining unit's voltage increases such that sufficient line voltage is present on the cable to power all of the repeater amplifier modules in the system. Figure 2.16 shows a sketch of nominal and failure PFE operation. When modeling PFE equipment for availability analysis, the performance can be represented by a simple redundant combination of components. Failure of both PFE units results in a system outage while an individual failure on either end of the link is not service affecting.

2.2.4 Submarine Fiber-Optic Cabling

Submarine fiber-optic cables are constructed of a number of different armor types. The armor is selected based on the seafloor survey and the region of installation. Harbors, ports, and beach landings typically use the heaviest armor while deepwater seafloor installations utilize light armor or no armor at all. Armor decreases the probability of a cable break due to fishing gear, anchor snags, and rocky seafloor wear. Reliability and availability modeling of submarine fiber-optic cabling is very challenging. Empirical data collection and analysis is the most accurate method of modeling. Detailed empirical analyses of fiber-optic cables installed in waters of similar profile and activity can be used to produce an aggregate cable model. Table 2.3 shows an example tabulation of empirical data for a number of submarine cables.

Using the empirical tabulation in Table 2.3, we can calculate an aggregate, average failure rate for the submarine cables in the region of interest.

$$L_{\text{total}} = L_1 + L_2 + L_3 + L_4 + L_5 = 11,800 \, \text{km}$$
$$T_{\text{total}} = T_1 + T_2 + T_3 + T_4 + T_5 = 376,680 \, \text{h}$$
$$f_{\text{total}} = f_1 + f_2 + f_3 + f_4 + f_5 = 8 \, \text{failures}$$

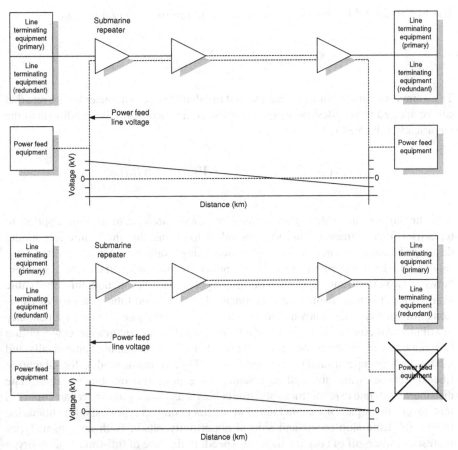

Figure 2.16. Power feed equipment operation, nominal and failure.

By dividing the total failure count by the total system operational time, we can find the failure rate for submarine cables:

$$z_{\text{submarine}} = \frac{f_{\text{total}}}{T_{\text{total}}} = 2.1 \times 10^{-5} \text{ failures/h}$$

Table 2.3. Submarine Cable Failure Rate Empirical Tabulation

Submarine Cable	Cable Length (km)	System Life (Years)	Failures Accumulated
1	1500	14	3
2	2000	7	0
3	800	2	2
4	3000	8	1
5	4500	12	2

The MTBF for submarine cables is given by the inverse of the failure rate

$$\text{MTBF}_{\text{submarine}} = \frac{1}{z_{\text{submarine}}} = 47085\,\text{h}$$

The submarine cable failure rate can be used to calculate a per kilometer failure rate that can be applied to new designs where the cable is installed in similar conditions to the empirically tabulated cables.

$$z_{\text{per km}} = \frac{z_{\text{submarine}}}{L_{\text{total}}} = 1.8 \times 10^{-9}\,\text{failures/h/km}$$

The failure rate calculation method presented above can also be applied to terrestrial fiber segments. Care must be taken to ensure that the failure rate model developed is appropriate for the application being evaluated. A metropolitan fiber segment model, for example, would likely not apply to a long-haul terrestrial segment between two cities running along a railroad right of way. When modeling fiber cable availability, it is reasonable to use exponentially distributed time-to-failure distributions for steady-state analysis. Recall that the failure rate of an exponentially distributed random variable is constant. The steady-state submarine cable failure rate can be assumed to be constant as the number of cables tabulated empirically and the total system operational time become large. Time-to-repair models for submarine fiber-optic cables are not well represented by exponential random variables. The distribution of time to repair for submarine cables is dictated by marine repair contracts that govern the repair of a cable fault. Two distinct contract types exist for submarine repair: full-time ship repair and ship-of-opportunity repair. Both agreement types represent a trade-off of cost versus repair speed. In the case of full-time availability, a ship is held in port on standby for repair at any time. This repair model results in the minimum time to repair with relatively low variability but commands the highest cost. Ship-of-opportunity repair agreements typically do not ensure that a repair ship will be on site within a given period of time. Transit times for ships of opportunity already performing other work can result in variable time-to-repair period. Repair distributions for submarine cables are best represented by normal or lognormal distributions with a mean and standard deviation representing the average time to repair and the variability of that repair. Figure 2.17 shows the PDF of a time to repair model representing a ship of opportunity with a mean value of 14 days and a standard deviation of 2 days. Thus, 95% of all ship repairs will occur within $10\,\text{days} \leq \text{TTR} \leq 18$ days. Properly written contracts will often ensure that these criteria are met as stiff penalties often follow when repair performance criteria are not met.

Availability analysis of submarine cables necessarily requires that Monte Carlo methods be used to produce results that are representative of actual performance. Use of Markov chain methods assume exponentially distributed time to repair and will result in erroneous availability results. Reliability analyses may utilize reliability block diagram analysis when the failure distribution is assumed to be exponential.

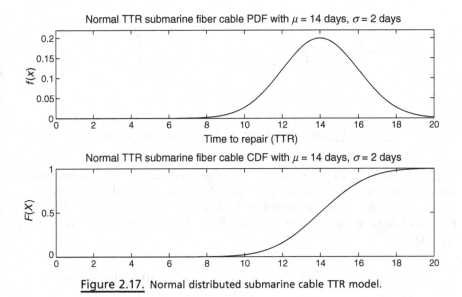

Figure 2.17. Normal distributed submarine cable TTR model.

2.2.5 Submarine Repeater Equipment

Long-haul fiber-optic systems require periodic amplification or regeneration in order to ensure that sufficient energy is present at the receiver. Submarine systems utilize fiber-optic repeaters to amplify the signal as it propagates along the fiber path. Fiber-optic repeaters represent one of the most critical points of failure within the system. Failures within the submarine plant result in long outages and expensive repairs. Implementation of redundancy within fiber repeaters is often not cost effective and for this reason, the fiber-optic repeater design focuses on producing hardware that does not fail. Many submarine fiber systems have multiple EDFA modules (typically for each fiber). In some cases, the repeaters are designed to accommodate the loss of one or more EDFA modules over the system lifetime. Although it is impossible to produce hardware that is failure free, the failure rate is made so low that the probability of a failure occurrence within the system lifetime is extremely low. Repeater failures rarely represent a significant number of lifetime system outages. Fiber-optic cable cuts and faults are far more frequent.

Calculations involving fiber-optic cable repeaters should focus on both reliability and availability. When determining system lifetime performance, it is of interest to calculate both the availability and the probability that the system survives its design life without failure (life-cycle reliability). Consider a submarine system with 10 submarine repeaters placed at equal periodic intervals of 80 km. The system described below in Figure 2.18 consists of 10 repeaters (without redundancy) that make up a serial combination of components.

Assume that all of the submarine repeaters are identical and have an identical failure rate equal to 5 FITs. The repeater unit consists of two discrete modules, an

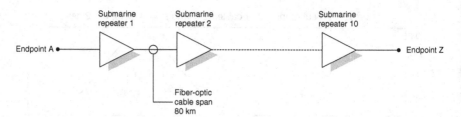

Figure 2.18. Sample submarine system with 10 periodic repeaters.

amplifier module, and a power module. The amplifier module is typically implemented using erbium-doped fiber technology. Figure 2.19 shows the RBD for a repeater.

If we assume that the amplifier module has a failure rate of 3 FITs and the power module has a failure rate of 2 FITs, then the total failure rate of each repeater is

$$\lambda_{\text{repeater}} = \lambda_{\text{amp}} + \lambda_{\text{power}} = \frac{3}{10^9} + \frac{2}{10^9} = 5 \times 10^{-9} \text{ failures/h}$$

Recall from above that the calculated failure rate of the submarine cable (due to external aggression) was

$$\lambda_{\text{per km}} = 1.8 \times 10^{-9} \text{ failures/h/km}$$

Assume that the submarine cable is designed for a system life of 20 years. Applying the definition of reliability and assuming that the failure rates of the repeater and the submarine cable are both constant (exponentially distributed):

$$R(t) = 1 - F(t)$$

where $F(t)$ is the cumulative distribution function of the failure distribution. In this case, the CDF is given by

$$F(t) = \begin{cases} 1 - e^{-\lambda t} & t \geq 0 \\ 0 & t < 0 \end{cases}$$

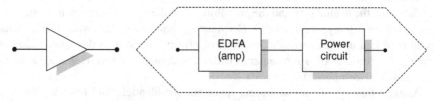

Figure 2.19. Submarine repeater RBD.

Thus, the reliability (survivor) function is

$$R(t) = 1 - F(t) = 1 - \left(1 - e^{-\lambda t}\right) = e^{-\lambda t} \quad \text{for } t \geq 0$$

The reliability of a serial combination of repeaters can be written as the product of the individual reliability functions:

$$R_{sys}(t) = R_1(t) \times R_2(t) \times \ldots R_{10}(t) = e^{-(\lambda_1 + \lambda_2 + \ldots \lambda_{10})t} = e^{-(50 \times 10^{-9})t}$$

The probability that the system survives 20 years ($T_{sys} = 175{,}200$ h) of operation without failure assuming only repeater failures is thus

$$R_{sys}\left(T_{sys}\right) = e^{-(50 \times 10^{-9})(T_{sys})} = 99.1\%$$

If we include the total fiber span length in addition to the submarine repeaters in the reliability calculation, we find

$$R_{sys}\left(T_{sys}\right) = e^{-(50 \times 10^{-9}) + (1.58 \times 10^{-6})(T_{sys})} = 75.2\%$$

Clearly, the failure rate of the submarine cable dominates the overall system reliability. Simply calculating the MTBF for each system provides far less insight into the system performance:

$$\text{MTTF}_{repeaters} = \frac{1}{\lambda_{repeaters}} = 20 \times 10^6 \text{ h}$$

$$\text{MTTF}_{system} = \frac{1}{\lambda_{system}} = 613 \times 10^3 \text{ h}$$

It is clear that the system is on a whole far less reliable than the submarine repeater subsystem, however, it is not clear how much less reliable it is. By calculating the probability of successful operation for the system life, it is easy to see a quantitative representation of this performance implication.

QUESTIONS

2.1. What are the primary sources of terrestrial fiber-optic cable failures?

2.2. A telecommunications company operates an installed base of 1000 km of terrestrial fiber-optic cable between two regions. Assuming the empirical failure data shown in the table below, calculate the failure rate for the cable on a per kilometer basis. Calculate the MTTR of a cable break and indicate that value as well.

Number	Date	Outage Duration (h)
1	1/30/2000	6.5
2	3/4/2001	14.1
3	9/3/2001	9.3
4	4/30/2002	11.3
5	5/14/2002	19.7
6	12/1/2002	3.4
7	2/18/2003	17.3
8	7/4/2004	7.4

2.3. What is the average availability of the fiber cable described in Q2.2?

2.4. Generate a best-fit Weibull distributed random variable for the TTF and TTR and plot the resultant PDFs for the data in Q2.2. Provide the scale and shape parameter values for each distribution.

2.5. Develop a Monte Carlo simulation for the Weibull distributions developed in Q2.4 for the fiber segments. Compute the fiber availability, TTF, and TTR for a 20-year life cycle.

2.6. A submarine fiber-optic cable is constructed between points A and Z and traverses 750 km of sea floor. The system utilizes nine repeater modules and has fully redundant PFE and SLTE equipment compliments. If each repeater module has a FIT rate of 18 FITs, calculate the predicted average availability of the cable assuming a failure rate of 6.7×10^{-8} failures/h/km and MTTR of 288 h for the submarine cable.

2.7. If a second submarine cable (identical to the cable in Q2.6) is installed on a diverse path using diverse landing station facilities and diverse PFE and SLTE hardware, what is the new achievable availability?

2.8. A vendor provides technical data claiming a submarine system availability of 99.99%. The data provided are listed in the table below along with a system block diagram. What MTTR assumptions must be made to achieve this target if the cable fault and repair impact is neglected? If the cable MTTF and MTTR are assumed as in Q2.6, what availability target is reasonable to expect for this system? Assume that the total submarine fiber length is 635 km and that there are seven repeaters along the cable span.

Component	MTBF (h)
SLTE (redundant)	220,000
PFE (redundant)	300,000
MUX (redundant)	180,000
Submarine repeater	5.6×10^7

2.9. Evaluate the availability impact of sparing submarine system components (for problem Q2.9) on site versus using a centralized depot. Assume that centralizing the spares increases the mean downtime by 18 h at both sites.

3

MICROWAVE NETWORKS

Terrestrial microwave networks are ubiquitous in modern communications. Microwave communications are used for a broad range of service delivery, from local area networks (such as Wi-Fi, commonly found in personal computers and electronics) to long-haul high-speed wide area networks (WANs). This section presents theory, considerations, and techniques for calculating reliability and availability of terrestrial microwave networks. Terrestrial microwave networks are split into a number of different types, each consisting of a number of fundamental components. Network types are:

- long-haul microwave
- short-haul microwave
- local area microwave

Long-haul microwave networks are typically deployed in environments and locations where fiber-optic cabling or coaxial copper cabling is very expensive or impractical. A long-haul network can be distinguished from short-haul or local area networks by the distances covered, frequency licensing, and multihop topologies. Long-haul microwave networks typically provide higher speeds and lower operating costs than satellite communications but have a higher initial investment. Satellite network

Telecommunications System Reliability Engineering, Theory, and Practice, Mark L. Ayers.
© 2012 by the Institute of Electrical and Electronics Engineers, Inc. Published 2012 by John Wiley & Sons, Inc.

latencies are much higher than those realized by microwave networks. Modern communication systems are designed to provide carrier-class performance with full equipment redundancy and high-availability path designs. Although infrastructure costs are generally lower than fiber-optic cables, the long-term expansion capability of microwave networks is more limited.

Short-haul microwave networks are generally designed with lower availability objectives than long-haul systems. They can serve as collector systems for long-haul systems or can provide in-region communication capabilities. Short-haul systems have a typical overall system route length equal to one tenth that of a long-haul system with the same availability objective. They can be multihop systems (typically operating in licensed frequency bands) or single-hop designs that may or may not operate in licensed frequency bands. Short-haul microwave links might be deployed in cases where fiber or copper cabling is not practical or to provide a redundant path in cases where the primary path is fiber or copper. Short-haul microwave networks have seen a recent surge in use in wireless cellular backhaul networks. High-bandwidth requirements of modern wireless handset data rates along with high-density cellular network topologies have driven increases in short-haul microwave network deployments. Reliability and availability analysis of short-haul microwave networks is often overlooked or ignored because of their lower cost (relative to long-haul network designs). Although the performance of off-the-shelf microwave components may be sufficient using standard design techniques, it can still be useful and informative to perform calculations to set expectations for customers regarding achievable performance.

Local area microwave networks are networks designed to generally provide limited coverage and high throughput. Common network technologies include Wi-Fi, Bluetooth, and WiMAX. Local area microwave networks are deployed in large quantities, reusing frequency spectrum across large geographic areas. Carrier-level production Wi-Fi network deployments ("Wi-Fi Hot Spot networks") can consist of hundreds or thousands of access point (AP) devices with millions of clients. The scale of such network deployments necessitates careful availability and reliability analysis to ensure that both network performance and maintainability achieve design targets. Equipment failures and truck rolls can quickly overwhelm operational expense budgets when underlying design targets and overlooking or ignoring analyses.

3.1 LONG-HAUL MICROWAVE NETWORKS

Long-haul microwave networks are very similar in topology and design to long-haul fiber-optic networks. The most significant distinction between microwave and fiber-optic long-haul networks is the signal path. Fiber-optic networks rely on a strand of fiber-optic glass to guide the signal along a path. This guided signal path somewhat simplifies the network design. Unavailability due to fiber-optic failures is dependent on continuity of the fiber-optic path. Failures of the path are most frequently due to vandalism, construction, or vehicular accidents. Microwave networks rely on an unguided over-the-air propagation path for the signal. This unguided signal path introduces a new source of signal unavailability in microwave networks. Over-the-air signal propagation is affected by a number of different physical quantities. Atmospheric effects, rain, reflective terrain and structures, and obstructions can all interfere with or block signals completely. Propagation availability can constitute a limiting factor in long-haul microwave network designs. Analysis of long-haul microwave network availability and reliability requires analysis of a number of discrete network components.

- Network equipment
- Signal propagation
- System topology

Although the analysis of long-haul microwave networks follows the same general procedure as long-haul fiber networks, the equipment, signal path, and network topologies are different. Specifically, the infrastructure supporting microwave antennas and repeater sites can have a major impact on system performance. Microwave towers and equipment shelters are often subjected to extreme weather conditions in remote locations. Network designs neglecting to address these conditions can result in catastrophic failures and long outages with expensive repairs.

Long-haul networks are implemented using either linear multihop network topologies or ring topologies. Long-haul microwave topologies are typically implemented as fiber overlay networks using the microwave links in place of traditional fiber-optic cables. Fiber-optic interfaces at OC-3, OC-12, or higher line rates connect to microwave radio baseband interfaces for transmission over the air. Figure 3.1 shows an example of a long-haul microwave network tower installed in western Alaska. This location is

Figure 3.1. Long-haul microwave network tower in western Alaska.

subjected to extreme weather conditions and operates using continuously running on-site generator power.

The similarities between fiber-optic and microwave long-haul networks intuitively lead to the conclusion that similar network availability and reliability performance is possible. Similar performance is achievable in long-haul microwave networks but it requires expert knowledge and attention to details that are specific to microwave.

3.1.1 Long-Haul Microwave Propagation

The propagation of electromagnetic waves is fundamental to all modern communications technologies. Long-haul microwave signal propagation has been studied by physicists and engineers for decades. Many books have been written on microwave design, propagation, and theory. Signal transmission calculations rely on empirical data collection and analysis. This section is intended to provide readers with a qualitative understanding of microwave propagation theory. Actual calculation of microwave path availability is beyond the scope of this book and is left for further study by the reader. The impacts and techniques for incorporating microwave path availability analysis into a microwave system design are within the scope of the discussion presented here. As such, it is relevant to provide enough background and theory to substantiate the models presented.

Microwave signals are transmitted by amplification of small signals by transmitter equipment. The amplified microwave signals are coupled to antennas that focus the signal energy in the direction of a receiver. This section examines the microwave propagation phenomenon and the factors that can affect availability and reliability performance. Atmospheric effects, rain, and terrain can all cause service impacting impairments in microwave systems. Design choices such as frequency diversity, antenna diversity, and adaptive coding and modulation technology can all be used to mitigate microwave transmission impairments.

3.1.1.1 Impairment Causes.

At the most basic level, microwave transmission impairments can be placed into two distinct categories: *multipath* fading and *rain* fading. Multipath fading is the variation of received signal energy due to reflections or refractions of the propagating wave arriving at the receiver after a time delay (with respect to the straight line propagating wave). Multipath comes in two forms: atmospheric and terrain-based specular reflections. The microwave path engineer can often mitigate the effects of specular reflection multipath. Use of appropriate repeater and terminal siting to block the reflected signal, frequency or space diversity, and antenna pattern discrimination can mitigate the effects of multipath. Both forms of multipath are frequently observed when signals propagate over large smooth areas such as bodies of water. Refraction occurs in the atmosphere when microwave signals propagate through stratified atmospheric layers causing an effective "bending" of the propagating wave. Sharp gradients in the atmospheric radio refractivity can form boundaries that cause atmospheric multipath. Experimental evidence shows that per hop (in multihop systems) multipath fading is uncorrelated and thus the total path unavailability due to multipath fading is the sum of the individual hop unavailabilities for the multihop

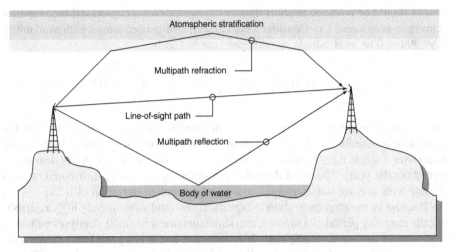

Figure 3.2. Multipath signal propagation.

system. Figure 3.2 shows a graphical representation of multipath in a microwave transmission system.

The total multipath fade outage experienced by a link is a statistical quantity and is influenced by a number of contributing factors. Various predictive models exist for evaluating multipath fading. Multipath fading that occurs due to reflective terrain is more deterministic in nature. Microwave engineers can often predict reflective terrain multipath through detailed analysis. Techniques are then applied to mitigate or limit the magnitude of the fading. It is the objective that most of this kind of multipath outage be eliminated through careful path design. The outage due to reflective terrain multipath is frequently not included in the calculation of propagation unavailability for microwave systems. The occurrence of atmospheric multipath is usually calculated based on historical empirical data compiled by experts who have studied many paths and modeled its occurrence. It is commonly mitigated by increasing the nominal received signal strengths (increasing the fade margin), use of space diversity (using two receive antennas spaced vertically), and/or use of frequency diversity (two parallel radio channels operating on different frequencies).

Multipath fading analysis focuses on characterizing unavailability due to short-duration outages that are caused by multipath events. It is important to note that many microwave path availability models define unavailability using the International Telecommunications Union (ITU) definition where 10 consecutive severely errored seconds (SES) are required to define a service as "unavailable." In order for a service to be considered "available," 10 consecutive error-free seconds must be observed. This definition of availability may or may not be applicable to the overall system model. In many cases, the hysteresis introduced by this definition of availability may obscure the actual performance of highly available systems. Outages due to multipath fading are typically characterized by hundreds of discrete events (with an average duration of 1 s) per year. When modeling multipath fading in microwave system availability models, the outages should be modeled as short-duration, frequent events (which self-recover after an average duration of 1 s). Consider a system with a calculated annual path availability of 99.999%. The total outage seconds per year would be

$$T_{\text{outage}} = 8760\,\frac{\text{h}}{\text{year}} \times 3600\,\frac{\text{s}}{\text{h}} \times (1 - 0.99999) = 315\,\text{s}$$

If the average multipath event lasts 1 s, a simple model might uniformly distribute the events across a single year. Figure 3.3 shows a graphical representation of 26 multipath events over a single month (assuming that the 315 s of outage are uniformly distributed throughout the year). The event duration is modeled as a normally distributed random variable with average duration equal to 1 s and a standard deviation of 0.25 s.

Because of the extremely short outage durations (and subsequently high availability), the sampling period of a Monte Carlo simulation must be small. Analyses with very small sample periods are computationally intensive and it is a good idea to determine whether a multipath fading analysis will result in a significant impact on the total system availability. As will be shown in future sections, rain fade, power systems, and hardware failure often contribute much more significantly to the total system unavailability.

Rain fade effects on microwave path propagation are characterized by long-duration outages due to interactions between the propagating wave and water droplets in the air. As the intensity (and the volume) of rain increases, the size of the water

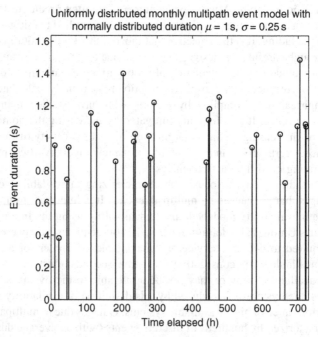

Uniformly distributed monthly multipath event model with normally distributed duration $\mu = 1\,\text{s}$, $\sigma = 0.25\,\text{s}$

Figure 3.3. Multipath outage event model using uniform occurrence distribution.

droplets increases. Microcell storms such as thunderstorms and squalls are often limited to a small geographic region but can have major effects on signal transmission. Rain fade varies with the transmission frequency. In general, lower transmission frequencies are less susceptible to rain fade with a maximum occurring at approximately 23 GHz. A number of rain models exist for modeling the rain behavior for a given region. Of these models, two have gained widespread acceptance in the microwave engineering world. The Crane model and the ITU-R P.530 are commonly used to model the unavailability of microwave radio paths due to rain fading events. When modeling rain fade effects on microwave radio links, the analyst must study the relative frequency of high-intensity rain events. Figure 3.4 shows a comparison of two different rain regions and their associated fade depth (in decibels). The first panel shows a region experiencing high rain clouds resulting in relatively low-intensity rain over a large region with a longer duration of occurrence.

The second panel shows a region that experiences thunderstorm activity over small geographic regions with a much shorter mean duration. The dotted line in both panels represents the fade margin of the link. The events in the second panel are much more likely to cause service affecting outages because the depth of the fade is much larger and thus exceeds the fade margin for a larger portion of the event time.

Many technological solutions exist for mitigating multipath and rain fade effects. These solutions provide the microwave engineer with the tools required to achieve specific availability targets on a path. The cumulative outage experienced by a microwave link is the additive combination of multipath (short duration, frequent events) and rain (longer duration, less frequent events) fade. Engineering designs incorporating diversity antennas, adaptive coding and modulation (ACM), and automatic power control (APC) need not be

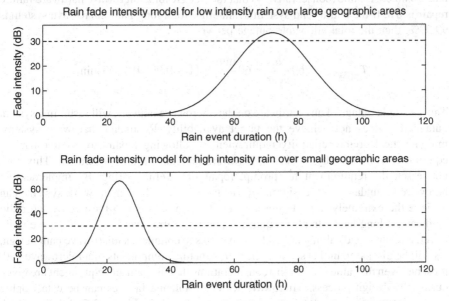

Figure 3.4. Multihop microwave radio link in a low-intensity rain region.

directly modeled in microwave transmission availability models. For example, the use of ACM and APC has the practical effect of increasing the fade margin of the system as long as the appropriate service assumptions are maintained. The overall link availability resulting from the engineering design is therefore sufficient to characterize the system's availability performance. Analysts must be careful to properly represent the duration and frequency of outages in propagation models. Analysts should strive to understand the characteristics of the fading events. Multipath events are usually unidirectional in nature and the outage estimated must be doubled to obtain bidirectional outage estimates. Rain fades, on the other hand, are bidirectional to the extent the fade margin is essentially the same in both directions of transmission (which is usually the case).

3.1.2 Long-Haul Terrestrial Microwave Equipment

Long-haul microwave communications equipment is designed to meet *carrier-class* telecommunications specifications. Long-haul microwave equipment is generally designed with the same level of redundancy as long-haul fiber-optic networks and central office switching equipment. Availabilities of 99.95%, 99.99%, 99.999%, or higher are commonly specified during the microwave design process. Achievement of this level of availability is only possible if hardware has a sufficiently low failure rate or is configured with suitable levels of hardware redundancy. In either case, the probability of an outage due to hardware failures must be determined to be negligible. Most long-haul microwave systems consist of one or more remote repeater sites. Microwave systems are most frequently deployed where high-speed fiber-optic cabling is not practical and the intermediate hop locations are often extremely remote and inaccessible. Assume (conservatively) that the MTTR for a microwave hardware failure repair is 8 h. If the end-to-end system availability for a long-haul microwave system is 99.99%, then the total allowable outage per year is

$$T_{\text{outage}} = 8760 \frac{\text{h}}{\text{year}} \times 60 \frac{\text{min}}{\text{h}} \times (1 - 0.9999) = 53 \, \text{min}$$

Thus, a single outage of any hardware on the microwave network will result in a system outage that does not achieve the target availability. Redundant hardware systems mitigate the extreme reliability requirement by allowing repairs to occur on failed equipment while standby or backup hardware maintains system operation. This does not lessen the requirement for prompt repair or careful sparing. By implementing hardware redundancy and designing repair models, long-haul microwave systems can achieve the extremely high design availabilities. In addition to the microwave radio hardware, a long-haul site also requires supporting infrastructure. A microwave radio tower, antennas, and cabling are all fundamentals to complete a microwave path design. As will be discussed in this section, the infrastructure components of a microwave path must be given the same availability considerations that the path and equipment are given during the design process. Microwave tower or antenna failures can be catastrophic service interruptions to a long-haul microwave network. Figure 3.5 shows a block

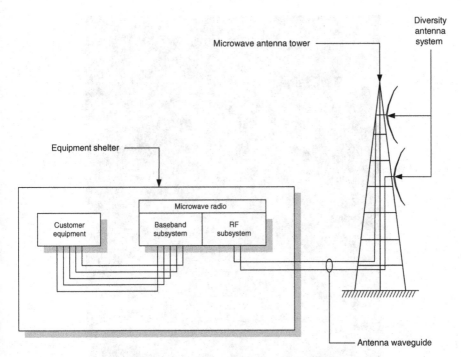

Figure 3.5. Long-haul microwave radio block diagram.

diagram of a long-haul microwave network radio including the redundant hardware and infrastructure components.

3.1.2.1 Infrastructure.
Line-of-sight microwave radio link design almost always requires some elevation of the radiating element (antenna) in order to couple sufficient energy to the radio receiver. Elevation of the radiating element is most often achieved by erection of a self-supporting, dedicated microwave radio tower. Microwave radio tower construction is a specialized engineering discipline requiring academic knowledge of structural engineering and practical deployment experience. Microwave towers and their foundations erected and constructed without sufficient attention paid to geography, soil type, load-bearing capacity, and climatic considerations can cause service impacting structural failures. Microwave tower failures can endanger the public and can cause long-term outages that are expensive and difficult to repair. Figures 3.6–3.8 show pictures of a microwave tower damaged by reoccurring ice formation. Note that although not easy to see, the tower diagonals in Figure 3.6 have been damaged due to falling ice.

Tower members and antennas were damaged due to ice buildup and shedding on the tower. Snow shields positioned above the antenna provide some protection from falling ice and debris. Fortunately, in this case, diversity antennas were deployed and switching equipment within the system architecture allowed the transmit communications path to be switched from the damaged antenna to the undamaged antenna. Diversity antennas

Figure 3.6. Microwave tower damaged by ice formation.

Figure 3.7. Ice bridge infrastructure damaged by ice formation.

Figure 3.8. Long-haul microwave antenna mount damaged by ice formation.

provide equipment redundancy in addition to improvement of the path availability performance. In remote and difficult-to-access locations, it is possible to use antenna diversity in combination with a remotely operable waveguide switch to manually or automatically reroute the transmitter to the diversity antenna even in cases where the path may not require diversity in order to achieve the required propagation availability.

Antenna waveguide is subjected to the same extreme climatic and environmental conditions as the antennas and towers. Care must be taken to ensure that waveguide runs are installed properly (and if high probability of icing is suspected, with ice shields for the horizontal runs) following certified tower procedures. Considerations for ice buildup, corrosion, humidity, and acts of God should all be considered when constructing long-haul microwave infrastructure. Each site location is unique and requires specific attention to ensure that failures will not occur that could cause extended outages.

3.1.2.2 Microwave Radio Equipment. Microwave antennas focus and couple the unguided microwave radio signal to a waveguide channel, ultimately delivering the signal energy to the microwave radio receiver. The microwave radio equipment consists of the hardware associated with the following functions:

- radio frequency (RF) signal transmission and reception
- system control, monitoring, and timing
- baseband interface transmission and reception

Figure 3.9 shows a block diagram of a sample long-haul microwave radio. This particular radio implements redundancy in all subsystems. This system also utilizes

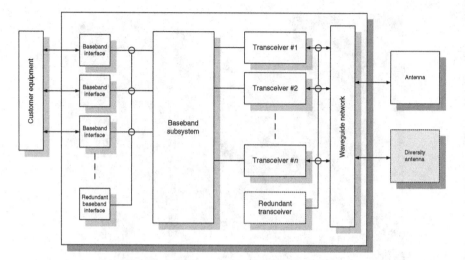

Figure 3.9. Sample microwave radio block diagram.

modern communications technologies such as cross-polarization interference cancel-
lation (XPIC) and adaptive coding and modulation (ACM). These technologies allow
the system to provide additional capacity and fade margin, enhancing system
performance.

The RF section of the microwave radio is made up of a group of transceivers that are
assigned to discrete frequency "channels" within the microwave radio frequency
spectrum. Long-haul microwave networks achieve high throughput by deploying a
number of channels operating at a relatively high line rate. For example, a long-haul
microwave radio might be initially deployed with $4 \times$ OC-3 (155 Mbps) frequency
diverse radio channels with the capacity to grow to $7 \times$ OC-3 channels in the future. The
transceiver elements connect to a waveguide network. This waveguide network acts as a
signal combiner on transmission and a signal splitter on reception. Transceivers often
offer an option of dual receivers to support designs where receive diversity is required.
Redundancy is implemented in a 1-for-N equipment protection design. In this design,
one complete standby transceiver protects N working transceivers. For the purposes of
reliability and availability analysis of long-haul microwave transceiver equipment, a
model should be developed that reflects 1-for-N equipment protection. Reliability
analysis of a 1-for-N system follows the techniques discussed in Section 1.3. Availability analysis should be performed using Monte Carlo simulation. The reason for
preferring Monte Carlo simulation is to ensure that the TTR model properly reflects the
operational plan of the microwave site. Long-haul microwave systems are often
deployed in circumstances where the TTR can have a large mean value and can
also have large variability. Normal, lognormal, or Weibull distributed random variables
are typically the best choice for simulating long-haul microwave TTR. Each system
design has its own geographic challenges and requires specific, careful attention to
ensure that the TTR assumptions properly reflect the expected system performance.

Consider a two-hop (three-site) long-haul microwave system that consists of two terminal nodes and an intermediate mountaintop back-to-back repeater. In this example, the microwave network consists of two frequency diverse traffic carrying channels backed up by a third frequency diverse channel. Figure 3.10 shows a sketch of this system.

We will calculate the reliability of the microwave transceiver subsystem for the entire microwave system for a number of different design lifetimes. Recall that the reliability is the probability that the system survives for a specified period of time under stated conditions. Assume that the MTTF of an individual transceiver (TRX) module is 125,000 h (as specified by the equipment manufacturer). In our analysis, we will assume that the TTF for these modules is exponentially distributed. Thus, the failure rate of the TRX module is

$$\lambda_{TRX} = \frac{1}{MTBF_{TRX}} = 8.0 \times 10^{-6} \text{ failures/h}$$

The reliability of an individual TRX module can be calculated using the definition of reliability

$$R_{TRX}(t) = 1 - F(t) = 1 - \left(1 - e^{-\lambda_{TRX}t}\right) = e^{-8.0\times10^{-6}t}$$

In this particular example, we will consider three scenarios. The first scenario is to calculate the system reliability using a single-thread TRX chain. In the second scenario, we will assume that the TRX's have one-for-one redundant modules in all cases. Finally, in the third case, we will assume that the transceiver system consists of two active channels at each location protected by a single redundant transceiver (one-for-two redundancy). Figures 3.11–3.13 show the RBD for each of the scenarios.

The reliability of the single-thread transceiver subsystem is

$$R_{TRX\_Sys}(t) = R_{TRX1}(t) \times R_{TRX2}(t)$$

The reliability of the one-for-one redundant transceiver subsystem is

$$R_{TRX\_Sys}(t) = 1 - (1 - R_{TRX1}(t))^2 \times 1 - (1 - R_{TRX2}(t))^2$$

The reliability of the one-for-two system can be calculated by applying the expression presented in Section 1.3:

$$Pr(S(X) \geq 2) = \sum_{y=2}^{3} \binom{3}{y} R_{TRX}{}^{y}(1 - R_{TRX})^{3-y}$$

$$Pr(S(X) \geq 2) = \binom{3}{2} R_{TRX}{}^{2}(1 - R_{TRX})^{1} + \binom{3}{3} R_{TRX}{}^{3}$$

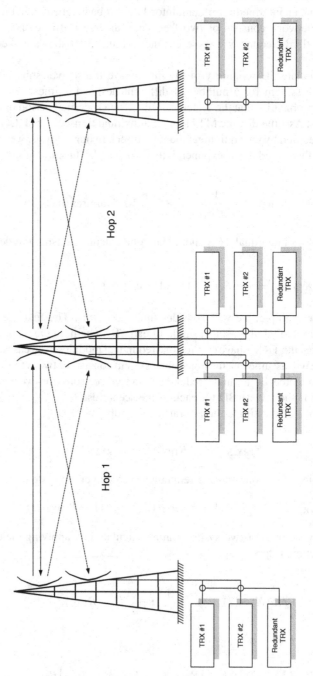

Figure 3.10. Two-hop radio transceiver system (one-for-two redundancy).

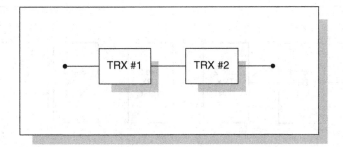

Figure 3.11. Single-thread transceiver system RBD.

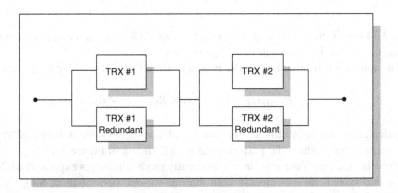

Figure 3.12. One-for-one redundant transceiver system RBD.

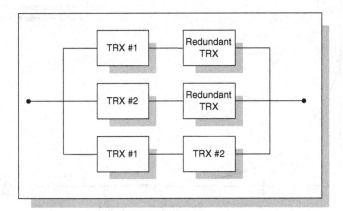

Figure 3.13. One-for-two redundant transceiver system RBD.

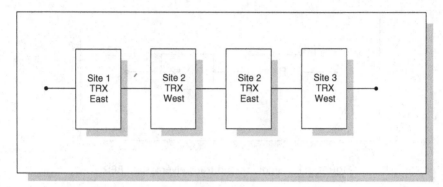

Figure 3.14. Two-hop radio link serial transceiver RBD.

Figure 3.14 shows the serial combination of the four TRX systems required to complete the two-hop radio link for each of the three scenarios.

The total system reliability is the product of the four TRX subsystem instances

$$R_{\text{sys}}(t) = R_{\text{Site1}_E}(t) \times R_{\text{Site2}_W}(t) \times R_{\text{Site2}_E}(t) \times R_{\text{Site3}_W}(t)$$

Calculating the reliability of the system of transceivers for a range of system lifetimes, we can produce the performance results shown in Figure 3.15 .

Note that the one-for-one system is inherently more reliable (as expected) while the one-for-two system follows closely with the one-for-one performance. Single-thread reliability follows well behind in reliability performance (also as expected). While the

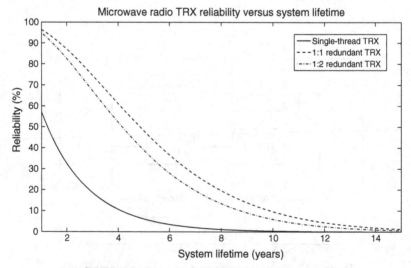

Figure 3.15. Microwave TRX path reliability comparison.

one-for-one and one-for-two systems have reliabilities that decrease somewhat linearly with increasing system lifetime, the single-thread system has an exponential decrease in reliability as system lifetime increases.

Some microwave radio systems implement control, management, and timing functions within each transceiver element while others implement discrete system-level components. In the case the control, management, and timing functions are implemented at the transceiver level, the analysis of system reliability and availability performance follows the same procedure as shown above (in these cases, the failure rate or MTBF includes transceiver functions as well as the control, management, and timing functions). In systems where these functions are implemented as discrete, independent components, a separate analysis is required to determine availability performance. In most cases, these functions are implemented using one-for-one redundant components. These one-for-one redundant system blocks would be placed in serial combination with the transceiver system block to determine the total system availability or reliability.

The long-haul microwave system also requires a baseband interface system block to allow other networks, customers, or end devices to send and receive network traffic over the microwave system. These baseband interfaces are sub-rate or line-rate interfaces that allow and enable access to the radio bearer (traffic channel) communications over the radio path. The baseband interfaces can be implemented using optical or electrical communications technologies. Baseband interfaces are almost always implemented as one-for-N redundant systems. Outboard multiplexing of signals to be transmitted is often performed prior to interfacing with the long-haul radio terminal. Figure 3.16

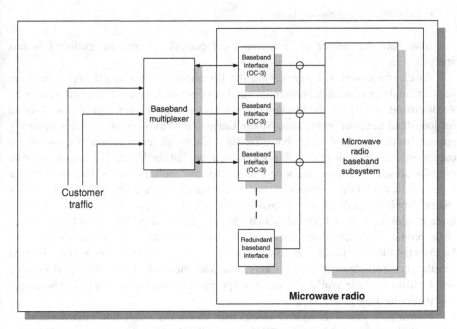

Figure 3.16. Long-haul microwave network multiplexed baseband OC-3 interface.

shows an example of a multiplexed OC-3 interface connecting to a long-haul microwave radio. If very high availability is the goal of a system design, it is important to ensure that not only the microwave radio transceiver, system control, and baseband functions but also any outboard functions (such as multiplexing or timing) are redundant.

3.1.2.3 Microwave Network Analysis.
Long-haul microwave radio networks are implemented in a variety of different topologies and almost always consist of a series of interconnected radio hops. Interfacing equipment such as SONET fiber-optic terminals, packet switched routers, and others allow complex network topologies to be deployed. This section will present the single-hop network design and analysis techniques as well as multihop and ring topologies. As network designs are becoming increasingly complex, the applicability of sample network designs diminishes. For that reason, this section will focus on a number of relatively simple designs in order to ensure that a clear representation of all system elements is provided.

Figure 3.17 shows a system block diagram of the elements involved in an availability analysis of a single-hop (of a multihop) long-haul microwave network where radio equipment redundancy is assumed to be one-for-one.

The major elements requiring analysis within this model are:

- outboard multiplexing equipment
- baseband radio equipment
- transceiver/RF radio equipment
- microwave path
- infrastructure considerations

Also note that power and environmental control systems are excluded in this analysis.

As can be observed by reviewing the list above, even a single-hop microwave network analysis contains a substantial number of network elements requiring analysis. As discussed previously, the Markov chain analysis method is not typically well suited for long-haul microwave analysis. Monte Carlo and Markov chain analyses generally require similar levels of effort but offer significantly different levels of detail in the output product. For this reason, it is recommended that the Monte Carlo simulation be used for all analyses involving long-haul microwave radio networks. In cases where a total system availability distribution is not necessarily required, the analysis of systems can be simplified to determine the availability of each serial network component. Once those availabilities have been calculated, the total system availability can be calculated as the product of the serialized elements. The danger with this type of analysis lies in the lack of visibility provided to network downtime. Network outage (or downtime) is often a critical contract element and it is rarely clear what the network downtime performance will be from a simple availability analysis (probability that the system is functioning at any particular time).

Figure 3.18 shows the system model rule set for the single-hop microwave radio Monte Carlo simulation. Note that TRX "A" and TRX "B" indicate the near and far end

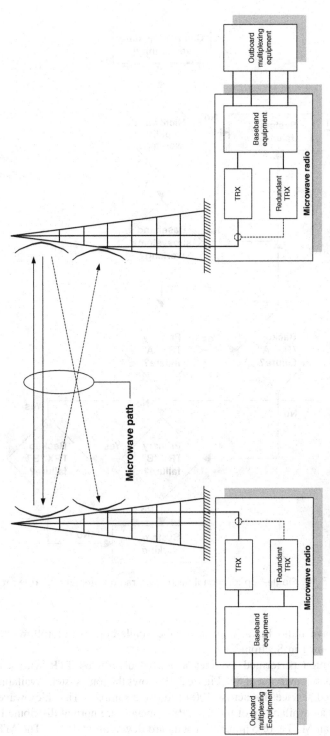

Figure 3.17. Single-hop long-haul microwave network block diagram.

113

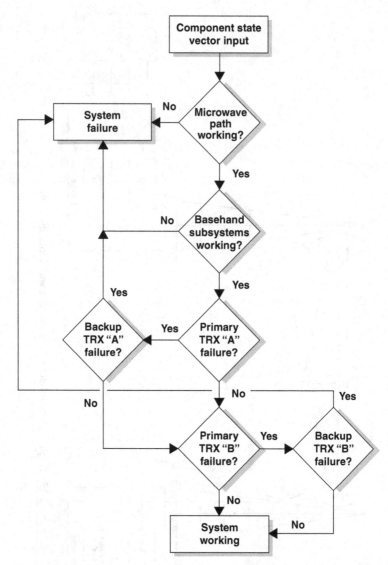

Figure 3.18. Single-hop long-haul microwave radio system model rule set.

of the microwave radio link. Analysis of the single-hop model follows the same procedure as shown in Section 3.1.

The simulation performed assumes a normal distributed TTR with a mean of 24 h and a standard deviation of 8 h. Figure 3.19 shows the total system availability for a 15-year designed service life across 1000 life-cycle samples. The microwave path is simulated with an availability of 99.99%. Path outages are normal distributed random variables with an MTTR of 5 min and a standard deviation of 1 min. The MTTR and

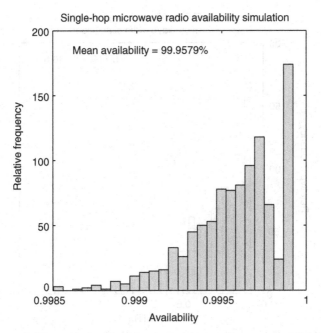

Figure 3.19. Single-hop long-haul microwave radio system availability.

standard deviation for the path-fading events is much shorter than the MTTR for the hardware-based components (baseband and TRX). The result of this system behavior is the "double-hump" exhibited in Figure 3.19. Hardware failures cause simulations to experience lower availability performance than simulation that experience only fading events in the system lifetime (represented by the large number of high-availability simulations in the histogram).

In addition to total system availability, the distribution of system downtime (outage) durations is shown in Figure 3.20. Figure 3.19 shows a distinct "double hump" in the observed availability distribution. In cases where the radio system does not experience a hardware failure, the availability will be very high (with low variability) because the only cause of unavailability in these cases is path impairments. In all other cases, the unavailability can be attributed to failures of either the TRX or the baseband systems. The variability in TTR in those cases led to much larger fluctuations in the overall achieved availability.

The behavior described above can also be seen in Figure 3.20 by noticing the distribution of TTR events where the large number of TTR events with very small values represent the modeled fading events and the rest of the distribution represents hardware-based repair.

An alternative analysis technique is to calculate the availability of each subsystem of the microwave link. The multiplicative product of these availabilities is calculated to be the total system availability. This approach simplifies the analysis and provides a

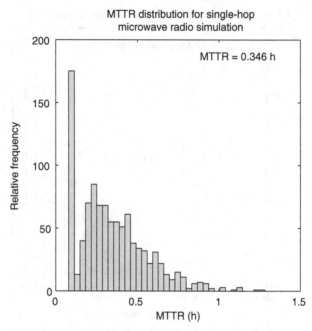

Figure 3.20. Single-hop long-haul microwave radio downtime distribution.

single predicted availability value as an output.

$$A_{\text{system}} = A_{\text{microwave path}} \times A_{\text{baseband}} \times A_{\text{TRX subsys}}$$

where A_{system} is the system availability, $A_{\text{microwave path}}$ is the calculated microwave path availability resulting from the microwave path study (which includes both multipath and rain fade characteristics), A_{baseband} is the availability of the baseband systems (including the outboard multiplexing equipment and the microwave radio baseband subsystem), and $A_{\text{TRX subsys}}$ is the availability of the one-for-one redundant transceiver subsystems. Each of these components can be analyzed using different, simplified analyses that are combined to provide an overall value. Although the results of this analysis are less detailed, the analysis itself is easier to perform and is often sufficient to satisfy design criteria. We can calculate the baseband subsystem availability as the product of the baseband availabilities at both ends of the microwave link.

$$A_{\text{baseband}} = A_{\text{radio } A \text{ BB}} \times A_{\text{radio } B \text{ BB}} = \left(\frac{\text{MTBF}_{\text{baseband}}}{\text{MTBF}_{\text{baseband}} + \text{MTTR}} \right)^2 = 99.968\%$$

The outboard multiplexing equipment availability is calculated in the same manner as

$$A_{\text{outboard muxing}} = A_{\text{outboard mux } A} \times A_{\text{outboard mux } B} = \left(\frac{\text{MTBF}_{\text{outboard mux}}}{\text{MTBF}_{\text{outboard mux}} + \text{MTTR}} \right)^2$$

where the equipment MTTR is 24 h. The transceiver subsystem requires a more sophisticated analysis technique (since redundancy is implemented). We will use the Markov chain technique here since we are focused on a simplified analysis (recognizing that the Markov chain technique requires that we assume an exponentially distributed TTR). Applying the hot-standby redundant availability model

$$A_{\text{TRX subsys}} = \left(\frac{\mu^2 + 3\lambda\mu}{2\lambda^2 + 3\lambda\mu + \mu^2}\right)^2 = 99.99996\%$$

where λ is the transceiver failure rate and μ is the repair rate

$$\lambda = \frac{1}{\text{MTBF}_{\text{TRX}}}, \quad \mu = \frac{1}{\text{MTTR}}$$

Thus, the total system availability can be approximated to be

$$A_{\text{system}} = A_{\text{microwave path}} \times A_{\text{baseband}} \times A_{\text{TRX subsys}}$$
$$A_{\text{system}} = 0.9999 \times 0.99968 \times 0.9999996 = 99.958\%$$

The single-valued average availability approach matches very well with the Monte Carlo simulation results. The advantage of performing the Monte Carlo simulation is that the results also include statistical distribution data for the TTR and availability that provides insight into performance variability as well as the expected value of system performance.

Analysis of multihop microwave networks follows the same procedure as shown above. Model complexity clearly increases with increasing topology complexity. The simplified model approach can be implemented for multihop network designs as long as the system does not use ring or mesh protection methodologies. Figure 3.21 shows a system block diagram for a network consisting of three cascaded instances of the single-hop network.

Determination of which analysis technique is appropriate is dependent of SLA requirements and customer deliverables. Ring or mesh network designs are the most complex and always require Monte Carlo simulation for accurate availability analysis. Mesh networks consisting of many nodes (and microwave paths) are challenging to analyze but these models can provide analysts, designers, and engineers with valuable insights into system weaknesses and design flaws. Sample analysis of mesh and ring networks is not presented here because of the highly variable, unique nature of every mesh network topology.

3.2 SHORT-HAUL MICROWAVE NETWORKS

Short-haul microwave networks are licensed and unlicensed microwave radio systems that are historically less than 250 miles in length with the same availability objective

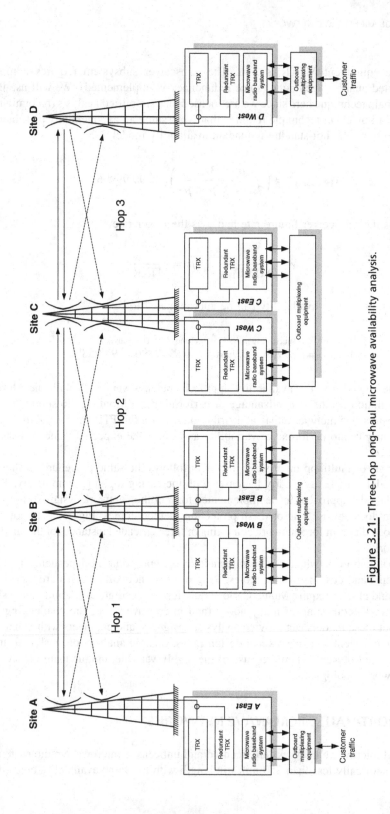

Figure 3.21. Three-hop long-haul microwave availability analysis.

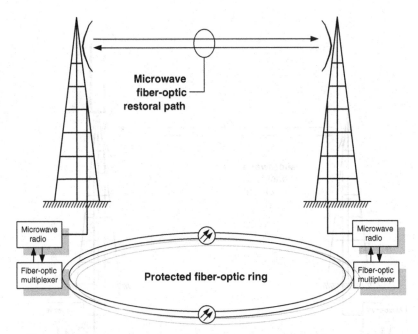

Figure 3.22. Short-haul microwave fiber optic ring network restoral path.

(99.98%) as a long-haul system 4000 miles in length. Although short-haul microwave radio was strictly defined in the past, modern deployments have seen short-haul microwave radio systems applied to a broad range of applications. These systems are often high capacity radio links and in many cases are designed to achieve high availabilities. Examples of some applications include redundant path recovery circuits for fiber-optic networks, cellular network backhaul links, and full-time network solutions. Figures 3.22–3.24 show sketches of some of these short-haul microwave network applications.

Short-haul microwave networks differ from long-haul networks in a number of significant ways.

- Path length is generally shorter
- Multipath propagation effects are less frequently experienced
- Equipment redundancy is less extensive

Cellular backhaul applications have driven technology improvements in the short-haul microwave radio market in recent years. Software-defined and packet radio technologies are now commonplace in short-haul radio designs.

3.2.1 Microwave Propagation

The physics of microwave electromagnetic propagation remain the same for both long- and short-haul microwave radio designs. Licensed short-haul microwave links in

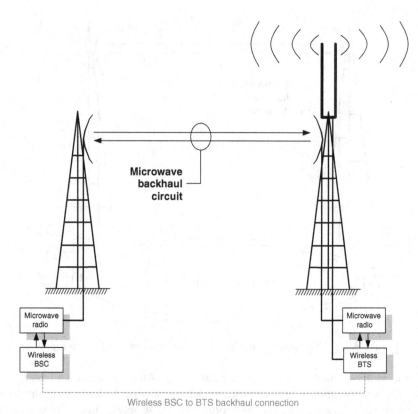

Figure 3.23. Short-haul microwave cellular network backhaul application.

particular are designed in exactly the same manner as long-haul links. For a complete treatment of reliability and availability implications of microwave radio propagation, the reader is directed to the previous section. One significant difference between long- and short-haul radio designs is with regard to the frequency of operation. While long-haul links are almost exclusively designed to operate in licensed frequency bands, short-haul links often operate within unlicensed frequency bands. This unlicensed frequency operation introduces another source of path unavailability, that is, interference. In the licensed band case, interference, although technically possible, occurs much less frequently since the process of licensing requires an interference analysis (in the United States), a prior coordination process with other incumbent users of the band in the area of the proposed new link, and the choice of frequency assignments that seeks to minimize the likelihood of interference with other links sharing the same frequency band. Unlicensed radio operation always runs the risk of performance degradations due to unforeseen interference. This interference can be a source of significant unavailability and link designers are well advised to utilize licensed frequency bands when feasible. Modeling interference in unlicensed bands is a very challenging problem due to the lack of data and difficulty in characterizing the nature of the interference. As an example, consider a link operating in the 2.4 GHz industrial, scientific, and medical

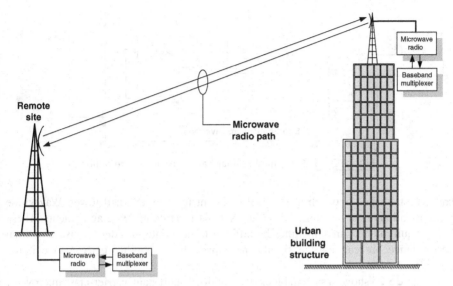

Figure 3.24. Short-haul microwave urban structure application.

(ISM) band. This frequency band is available for use by any transmission device conforming to a relatively basic set of equivalent isotropic radiated power (EIRP) restrictions. The decision to operate in this band automatically subjects the operator to unpredictable, potentially service affecting interference events that may or may not be resolvable. The ISM band is also shared by industrial, scientific, and medical equipment that can create significant amounts of radio interference (such as industrial microwave ovens).

All other aspects of short-haul microwave propagation remain identical (or extremely similar) to long-haul designs. The analysis of multipath and rain fade follows the same procedure, although in the short-haul case, both of these effects are usually less significant simply because of the fact that the link distance is somewhat shorter.

3.2.2 Microwave Equipment

All microwave networks rely upon a standard set of hardware required for link construction (as discussed in the long-haul microwave network section). Short-haul networks require antenna mounting infrastructure, antennas, transceivers, and baseband interfaces just as in the long-haul case. The differences lie in the manner in which each of these system elements is implemented. Infrastructure elements such as microwave towers, waveguide, and antennas are often more variable and depend on the site location and circumstances. For example, waveguide feed lines are eliminated when the radio terminal is mounted directly behind the antenna. In those cases, the radio is connected directly to the antenna feed. In the case of cellular backhaul, microwave radios are often deployed in urban environments where multistory buildings exist. Multistory buildings are frequently used to provide a "tower" structure so that line-of-sight communication

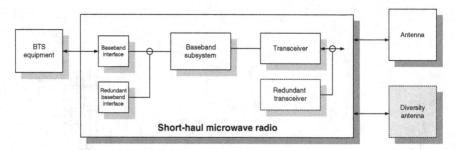

Figure 3.25. Short-haul cellular backhaul microwave radio.

can be established between a backhaul endpoint location and cell tower. Waveguide routing and antenna placement are often subject to building lease agreements. These lease requirements can sometimes be difficult to negotiate in order to ensure optimal performance but can have a significant impact on availability performance if not considered during the site selection and design process.

Figure 3.25 shows a system block diagram for a short-haul, carrier-class microwave radio designed for use in cellular backhaul applications.

Figure 3.26 shows a system block diagram for a single-thread, unlicensed short-haul microwave radio that might be used for low-capacity service delivery at the commercial level.

The redundant backhaul radio system is capable of achieving much higher availabilities than the unlicensed, commercial radio link due to its ability to utilize redundant baseband and transceiver equipment as well as diversity antennas.

3.2.3 Microwave Network Analysis

As an example, consider an analysis of a single-hop system for each of the two microwave radios shown in Figures 3.25 and 3.26. We will calculate the availability of the two systems using the simplified approach presented earlier in this chapter. This approach can be particularly useful when the relative performance of two platforms is of interest. In that case, the exponentially distributed TTR assumption "falls out" of the analysis since we are more interested in the difference in performance between the two systems rather than the absolute availability value. Both systems have the same

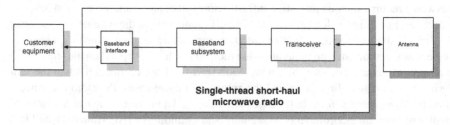

Figure 3.26. Unlicensed short-haul commercial service microwave radio.

Table 3.1. Short-Haul Microwave Availability Comparison Parameters

System Element	Value
Baseband equipment	150,000 h
Transceiver equipment	95,000 h
Path availability	99.95%

fundamental availability expression. The difference in the two lies in the fact that the single-thread system has a substantially lower upper limit on its achievable availability for the baseband and transceiver subsystems due to the lack of redundancy.

$$A_{\text{system}} = A_{\text{path}} \times A_{\text{baseband}}^2 \times A_{\text{TRX}}^2$$

where A_{TRX} and A_{baseband} are squared because there is an instance at each end of the link. The baseband and the TRX subsystem availabilities are determined using Markov chain analysis as in the long-haul case. Table 3.1 tabulates relevant parameters for calculating system performance.

Figure 3.27 shows curves for the achieved availability of each system at varying values of MTTR.

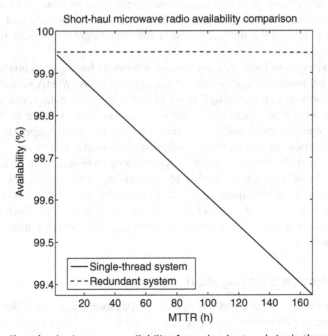

Figure 3.27. Short-haul microwave availability for redundant and single-thread designs at varying MTTR values.

Note that although the radio link propagation availabilities are designed with identical 99.95% target availabilities, the hardware in the single-thread case is non-redundant and results in a lower achieved availability. The availability in the redundant case is almost completely determined by the path availability since the hardware availability is an order of magnitude higher than the path availability. In addition to the hardware availability, limitations in the single-thread case, the difficult to quantify potential for interference, also exists.

3.3 LOCAL AREA MICROWAVE NETWORKS

Both long- and short-haul microwave networks are designed to deliver service between two fixed endpoints. This point-to-point network topology leads the network analyst to approach the availability and reliability in a similar manner. That is, the availability of the point-to-point networks is dependent on the probability that service will be available at any particular instant in time at both the A and Z endpoints of the network. Local area or multiuser networks require a somewhat different approach to availability analysis. In the local area network case, the failure of one network element may affect service to all users, some users, or no users at all. The availability is thus dependent on the number of users and number of network elements. Figure 3.28 provides a qualitative sketch of this concept.

Local area wireless networks are comprised of a number of different technology-dependent network components. These components are can generally be referred to as APs, customer premise equipment (CPE), backhaul equipment, and core equipment (in some cases). Figure 3.29 shows a generic example of a local area wireless network.

Some examples of local area microwave network technologies in production today are Wi-Fi, worldwide interoperability for microwave access (WiMAX), and Bluetooth. All of these networks allow multiple users to access the same radio resource through a multiplexing technique. Local area microwave networks can operate in licensed and unlicensed frequency bands and offer a range of different service types. Bluetooth is a technology focused on very small "personal area" networks with coverage regions on the order of 10 m. Wi-Fi increases the coverage area (typically 30 m or less) but still maintains a focus on a small number of concurrently connected users. WiMAX is designed to offer both ubiquitous (e.g., a large metropolitan area) access coverage (the range of coverage might be as much are 30 km in line-of-sight conditions, but is typically 10 km or less) and short-haul microwave radio functionality, although not both simultaneously.

3.3.1 Microwave Propagation

The nature of local area network microwave propagation requires a different analysis than that used in point-to-point propagation analysis. In local area wireless networks, the path availability is typically not explicitly calculated. Rather, in local area applications, regions are typically calculated that determine the boundaries of coverage.

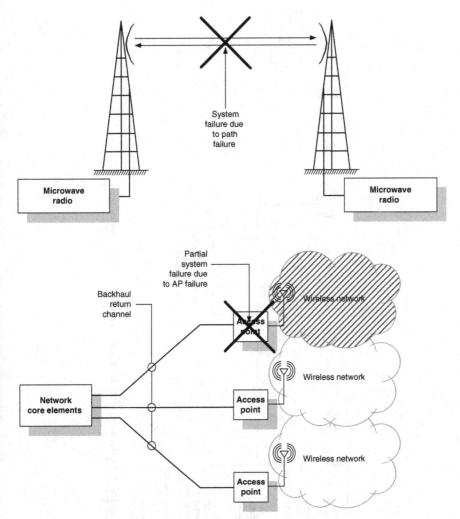

Figure 3.28. Point-to-point versus local area network topology failure modes.

End users operating within this coverage region are assumed to have available service. Figure 3.30 shows a sample "heat map" demonstrating the coverage region calculation.

When analyzing service availability for local area wireless networks, it can be difficult to predict the path availability of a particular end user. Mobility and obstructions (also call clutter) introduce interference that is difficult to quantify in terms of service availability. It is reasonable to select a threshold received signal level (RSL) that incorporates sufficient fade margin to accommodate both slow and fast fade events that are experienced in mobility applications. Users operating within the regions for which the desired RSL are achieved would be modeled as being available. Selection of the appropriate RSL is dependent on the technology and frequency of operation.

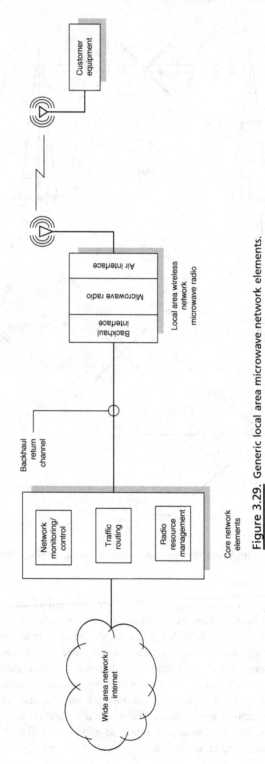

Figure 3.29. Generic local area microwave network elements.

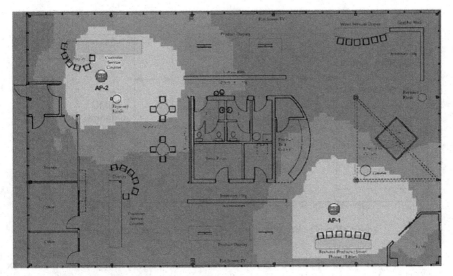

Figure 3.30. Local area wireless network heat map coverage region.

3.3.2 Microwave Equipment

Hardware deployed in conjunction with local area wireless networks is often offered at a much lower cost than the equipment used in long- and short-haul microwave networks. The lower cost of this equipment necessitates performance trade-offs. In the case of local area wireless networks, redundancy is rarely implemented at the access point level. Increased availability performance is usually achieved by deploying multiple over-lapping radio "sectors" that allow failures to be masked by offering additional capacity if one of the active devices were to fail. Supporting structures, antennas, and cabling are often deployed with a far less rigorous approach to engineering. Failure to address these fundamental aspects of microwave network design, even in the case of local area wireless networks, can lead to a lack of performance achievement in deployments. In the same vein as long- and short-haul microwave systems, it is important to ensure that antenna mounting, cabling (or waveguide), and support structures are all carefully considered to ensure that failures do not occur. Failures of infrastructure (and hardware) components in systems with hundreds (or thousands) of deployed units are both expensive and logistically difficult to support. Careful attention to the subtle details up front brings worthwhile operational cost savings and availability performance improvements throughout the system life. Local area wireless networks can be broken into two basic functional types:

- transceivers/access points
- core network elements

Core network elements are often found in medium to large local area wireless network deployments where many transceivers are deployed, such as large, multi-venue

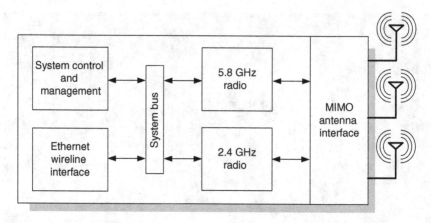

Figure 3.31. Wi-Fi access point functional block diagram.

"Wi-Fi Hot Spot" networks. The core network elements allow many subscribers to be serviced simultaneously while also providing network management, control, monitoring, and traffic routing functions. In small network deployments, the transceivers may have sufficient "intelligence" to operate autonomously. This can improve the system availability by reducing the number of deployed components but typically comes at the price of increased network management effort due to the less elegant design.

When modeling local area wireless network elements, the transceiver is typically specified by a MTBF value that represents the entire functional block. This functional block includes RF reception and transmission, system control/management/operation, and backhaul interfaces. Figure 3.31 shows a sketch of a sample Wi-Fi access point device that incorporates all of these elements of operation.

Failure of transceiver elements frequently involves replacement of the entire device (rather than parts of subsystems within the device). In some equipment types, the RF section and control section of the radio are separated using an outdoor unit (ODU) and an indoor unit (IDU) to improve RF performance (Figure 3.32). In these cases, the ODU and IDU might be serviced and/or replaced separately. The MTTR for ODU and IDU devices is often different due to the effort involved with the replacement of each device type.

In networks utilizing core elements for control, management, and reporting, it is common to utilize redundancy for the hardware. Large and medium network sizes (hundreds to thousands of deployed transceivers) carry significant network traffic and an outage of the core elements would result in significant unavailability network-wide.

3.3.3 Network Availability Analysis

Network analysis is focused on determining the availability of all network elements and ensuring that the network meets design criteria for availability and time to repair. As discussed previously, the core elements supporting local area wireless networks are often serving hundreds or thousands of network elements (access points). As such, these

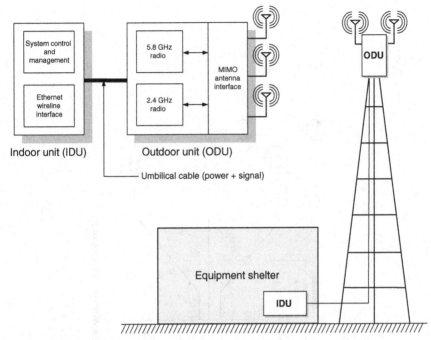

Figure 3.32. Radio design types, integrated versus split (ODU/IDU).

core elements represent critical network components and should be designed with criticality in mind. Figure 3.33 shows a network diagram of a local area wireless network implementing 802.11n Wi-Fi technology that consists of a large number of access point devices geographically distributed across a large area. The core network elements support subscriber registration, billing, management, and traffic routing. The access point devices provide subscribers with network edge wireless access. Backhaul network elements connect the core to the distributed access points using a variety of communications technologies.

In this example, we will calculate network availability and sparing requirements for the network. Network availability is defined as the aggregate availability of all network elements. This includes core elements, access point elements, and backhaul circuits supporting the delivery of traffic from the wireless coverage area to the core routing equipment. Subscriber availability might vary based on the type of backhaul utilized. If backhaul capacity is delivered using a variety of techniques, it is important to pay attention to the backhaul transport used to ensure that SLA requirements are met. Assume that the network elements within the core are implemented such that the core availability achieves a target of 99.995%. The routing/switching, authentication, and network management elements would all be implemented as redundant components in order to achieve an availability of this magnitude. Assume that the MTBF of an access point is 75,000 h and that the MTTR of access point repairs is 36 h. For the purpose of this example, we will consider a system in which 1500 access points are distributed

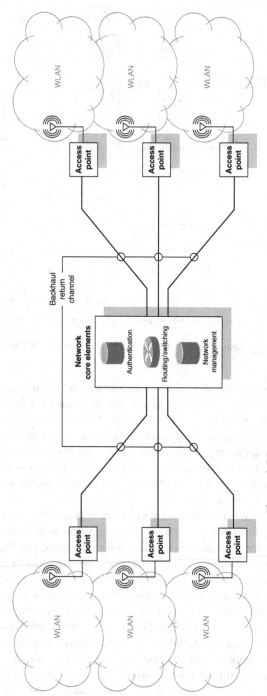

Figure 3.33. Sample Wi-Fi local area wireless network diagram.

across a large geographic region. The availability of the access points is (on average) equal to the generalized availability expression

$$A_{AP} = \frac{MTBF_{AP}}{MTBF_{AP} + MTTR} = \frac{75,000\,h}{75000\,h + 36\,h} = 99.95\%$$

This availability is independent of the number of deployed access points. We can calculate the required number of spare access points per year as

$$S = \frac{AP\,Count \times Annual\,Hours}{MTBF_{AP}} = \frac{1,500 \times 8,760\,h}{75,000\,h} = 175.2\,failures/year$$

where S is the required number of spare units per year. Although the achieved availability performance of 99.95% seems reasonable for the access points, in general, when the network scale is examined at 1500 units, the failure rate is clearly not reasonable. Since the sparing level scales linearly with increasing access point MTBF if we double or triple the MTBF for the access points, we will cut the spares requirement by half of 1/3 of the original value (annually). By determining the number of predicted spares consumed per year, we can also calculate expected operational repair expenses and budget accordingly.

Since the core availability is an order of magnitude greater than the access point availability (99.995% vs. 99.95%), the core availability does not contribute significantly to the system unavailability. Although that is true, a failure of the network core is catastrophic and affects all users and subscribers. Because of this sensitivity, spares should be kept on hand and configurations should be regularly backed up to ensure that system failures can be repaired quickly and effectively.

QUESTIONS

3.1. A single-hop microwave radio link is designed for a propagation availability of 99.995%. If the microwave radios on either end of the link have aggregate MTBF values of 165,000 h each and the MTTR of the system is 14 h, what is the predicted availability of this network?

3.2. Calculate the predicted number of failure for the system described in Q3.1 assuming a 20-year system life. What is the expected annual availability for the years in which a failure occurs?

3.3. Provide a description of why the availability and reliability impacts of microwave network infrastructure (such as antennas, towers, and others) are difficult to model in system analyses. What methods can engineers use to ensure that the performance of these components meets with design expectations?

3.4. Assuming that the system components described in Q3.1 are used to implement a four-hop microwave system (with back-to-back microwave radios at the intermediate sites), what is the resultant availability? What MTTR must be selected to achieve a system availability of 99.9% and 99.99%?

3.5. Compute the availability impact of a microwave system upgrade that adds six channels to an existing 1:2 system. The new system design implements 1:8 component redundancy. Assume that each microwave radio TRX has an MTBF of 85,000 h and the MTTR of a module is 36 h.

3.6. A fiber-optic ring network achieves an availability of 99.9% using a combination of equipment and path diversity. Calculate the impact of adding a short-haul microwave radio link for system redundancy. Assume that the availability of the microwave radio link is 99% and that the MTTR for both systems is 48 h.

3.7. A large-scale Wi-Fi network is being deployed with approximately 4500 units. If the vendor-provided MTBF is 165,000 h for an access point device, estimate the number of spares to be purchased for annual repair and replacement of failed units. What MTTR is required to achieve an availability of 99.8% and 99.99%?

4

SATELLITE NETWORKS

There are many regions of the planet where satellite technology provides communications services more economically than any other method. In many rural locations, satellite communications provides the only connection to the outside world in rural locations. The critical nature of this communications link requires a focus on availability and reliability during the design process. Satellite communications is utilized to deliver a variety of services ranging from telephone service, Internet connectivity, and television to telehealth and distance learning. Telehealth and education services frequently demand the highest network availability performance. Clinic customers rely on the availability of services to ensure that patients receive the highest quality of care. Modern technology has enabled telehealth systems to deliver a wide range of communications services including video conferencing, electronic records, and voice communications. Remote industrial applications, including mining and oil exploration, and environmental applications can place challenging performance demands on satellite networks. Industrial satellite network customers often require high reliability or availability performance to ensure continuity of business. Clear definition of availability targets is a very important step that ensures that the cost or performance trade-off is properly balanced.

This chapter discusses the aspects of satellite network design relevant to availability (and reliability) analysis. Propagation effects, although similar to terrestrial networks,

Telecommunications System Reliability Engineering, Theory, and Practice, Mark L. Ayers.
© 2012 by the Institute of Electrical and Electronics Engineers, Inc. Published 2012 by John Wiley & Sons, Inc.

have distinct differences in satellite applications. Frequency-dependent propagation impacts on availability are discussed for C-band, Ku-band, and Ka-band systems. Other frequencies of operation, less commonly encountered in commercial satellite systems, are also discussed. Satellite earth stations are divided into two types. The very-small-aperture terminal (VSAT) network discussion includes analysis techniques for antennas, low-noise block downconverters (LNBs), block upconverters (BUCs), and modems. VSAT networks most often operate at Ku-band and Ka-band frequencies and typically do not employ dedicated shelters for equipment. Remote earth stations and teleports are presented with discussions of frequency conversion equipment, transmitters, low-noise amplifiers, modems, and site monitor and control equipment. The impacts and importance of each of these network components is addressed. Relevant applications of redundant equipment are presented as well. Teleports and hubs focus on site availability and RF subsystem redundancy. Multiple antenna deployments, high-power amplification equipment, and carrier-monitoring methods are also discussed.

Spacecraft reliability represents a significant business risk to both satellite operators and customers. Customers must weigh the risk of satellite failure against the significant cost of backup capacity. Backup options and redundancy techniques available for spacecraft protection are discussed in detail.

The chapter closes with a discussion of satellite network topologies and reasonable performance expectations for each network type. Hub/remote and point-to-point networks are discussed in this section.

4.1 PROPAGATION

Satellite signal propagation affects the availability performance of satellite networks due to interactions between the atmosphere and the propagating electromagnetic wave. This effect (analogous to the effect observed in the terrestrial microwave environment) is dependent on the frequency of transmission and the environmental conditions present at the time of transmission. The signal attenuation that a particular link can accommodate is often referred to as fade margin. In addition to attenuating a signal, rain (and other atmospheric effects to a lesser extent) can cause depolarization interference. Although depolarization of the signal does not directly cause interference, the depolarization of other signals within the communications payload can cause interference on cross-polarized channels.

This section examines the frequency-dependent fading effects for common satellite transmission bands (C, Ku, Ka, X, and L) and the methods employed to calculate link availability as a result of these effects.

In addition to environmental fading effects, satellite propagation is sensitive to interference from a number of sources. Active interference management and mitigation is an ongoing effort and requires attention from spacecraft operators and customers. Although difficult to predict (and thus plan for), the impact of interference can be catastrophic to satellite link availability. Frequent (or continuous) performance measurement is an important element of satellite network design to ensure that interference effects are quickly identified and mitigated. These proactive performance management techniques ensure that satellite network customers achieve the highest possible network availability performance.

4.1.1 Frequency Bands of Operation

Satellite communications operate in a number of carefully governed frequency bands. The limited quantity of available electromagnetic spectrum places specific guidelines for the frequency of operation and the allowable power transmitted from any particular station. In the United States, satellite network licensing is governed by the Federal Communications Commission (FCC). Other countries (or conglomerates of countries) have similar commissions or governing bodies. Commercial satellite networks generally operate in one of three frequency bands (C-band, Ku-band, and Ka-band). In the United States, the government and military agencies have access to an expanded range of frequency bands for operation including X-band, S-band, and L-band. Table 4.1 enumerates the commonly used frequency bands, their primary user, and the specific frequencies of operation allowed. Many bands are not exclusively used by satellite networks and operators. Terrestrial services sharing spectrum must coordinate operations to mitigate interference between the services.

4.1.1.1 Rain Effects. Rain affects the performance of satellite networks in a frequency-selective manner. In general, satellite networks operating at lower frequencies of transmission and reception are less sensitive to rain and atmospheric water content. Systems operating at higher frequencies (Ku-band and Ka-band) can experience significant signal fade due to rain activity. The magnitude of the rain attenuation and depolarization is dependent on the rain intensity. Significant research and development has been done to both model and predict rain fade for satellite links. The technical details of modeling frequency-selective rain fade are beyond the scope of this book. The impact of these effects is fundamental to the design and operation of satellite communications links. Equation 4.1 describes the predicted attenuation (α) in dB/km due to rain based on the point rain rate R (in mm/h) and the frequency-dependent

Table 4.1. Satellite Network Frequencies of Operation

Frequency Band	Primary User	Frequency of Operation
C-band	Commercial	Downlink 3.7–4.2 GHz
		Uplink 5.925–6.425 GHz
Standard fixed satellite service (FSS) Ku-band	Commercial	Downlink 11.7–12.2 GHz
		Uplink 14.0–14.5 GHz
Ka-band	Commercial	27.0–40 GHz
X-band	Military	Downlink 7.25–7.75 GHz
		Uplink 7.9–8.4 GHz
S-band	Government	2.0–4.0 GHz
L-band	Military, government	1.0–2.0 GHz

attenuation coefficients a and b (Pritchard et al., 1993).

$$\alpha = aR^b \tag{4.1}$$

The specific attenuation α must be converted to a total attenuation based on the total slant range to the satellite (Pritchard et al., 1993).

$$A = \alpha L \tag{4.2}$$

where A is the total attenuation (in dB) and L is the slant range to the satellite (one way). The point rain rate is determined using tabulated rain data that specifies the rain "zone" based on the geographic location of the earth station location of interest. Attenuation coefficients are tabulated based on the frequency of operation and the polarization of interest (horizontal, vertical, or circular). Readers interested in the technical details of point rain rate or attenuation coefficients are referred to the CCIR standards body. The CCIR houses references providing substantiating detail for rain rate and attenuation coefficients. The predicted attenuation is calculated based on the probability that the point rain rate is exceeded for some percentage of the year. Thus, the fade margin required in the link calculation is based on both the rain rate and the probability of that rate being exceeded. The higher the confidence in the result, the more fade margin required. Thus, the availability of the satellite path can be calculated based on the probability that the designed link fade margin is not exceeded. The uplink and downlink path availabilities are incorporated into the overall network availability analysis.

4.1.1.2 Multipath Effects.

Multipath was discussed in the terrestrial microwave system section and refers to multiple reflected copies of a signal arriving at the receiver at different times. The arrival of multiple copies of the same signal waveform can result in both constructive and destructive interference effects. Destructive interference due to multipath propagation can cause significant fade in microwave transmissions. In the case of satellite communications networks, the angle of incidence (elevation angle) made between the ground station antenna and the satellite is typically quite large. This large elevation angle generally causes multipath effects to be minimal since the reflected wavefront (for large elevation angles) does not appear at the receiver due to the directivity of the ground station antenna (its receive band main beam pattern will discriminate and attenuate the ground-based reflected signal). Small elevation angles occur in satellite ground stations only at very high latitudes (polar or nearly polar locations). In those circumstances, the effects of multipath propagation may become significant enough to affect service and should be considered for those circumstances. Figure 4.1 shows a sketch of a ground station antenna at different earth latitudes and the associated elevation angle.

Recalling geometry and reflection/refraction criteria, we can calculate the minimum (or critical) angle for reflection of the waveform. Multipath reflection is dependent on the surface roughness and the frequency of operation. Very smooth regions (such as calm bodies of water) make good reflectors while other surfaces may "appear" smooth across a limited range of frequencies.

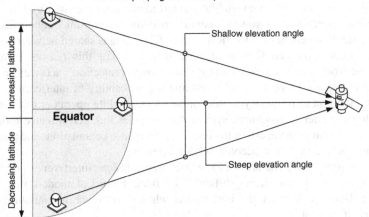

Satellite propagation multipath scenarios

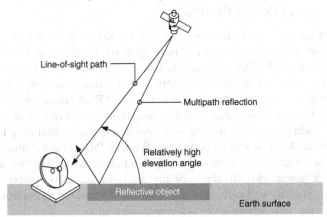

Steep elevation angle multipath mitigation

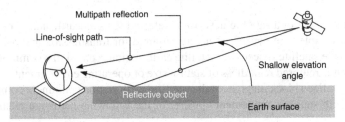

Shallow elevation angle multipath condition

Figure 4.1. Satellite earth station multipath condition sketch.

4.1.1.3 Interference Effects. Signal interference can represent a significant source of satellite network unavailability. Interference can be caused by terrestrial or space sources. Terrestrial microwave networks employ transmission frequencies that are shared by satellite networks. Specifically, the 6 GHz band is shared between long-haul terrestrial microwave and C-band satellite networks. For this reason, commercial satellite network carriers often employ "frequency protection" services to ensure that frequencies are properly coordinated and the probability of interference is minimized. In addition to frequency protection services, satellite operators and customers can deploy a carrier management system that provides link performance monitoring data. This data allows the operator to observe performance degradations and potentially take action before the event become service affecting.

The unpredictable nature of satellite transmission interference makes the unavailability due to its effects difficult to model. Empirical models that rely on historical data tend to offer the best results when interference unavailability is the parameter of interest.

4.1.2 Automatic Power Control

Many modem hardware manufacturers offer technology that allows the modem to control its uplink power based on the received signal level at the far end. Systems using automatic power control technology accomplish two optimizations of the satellite link. First, the link power is minimized and this minimization results in a reduction of interference effects on the satellite. Second, the required RMS power for transmission is greatly reduced. Rather than continuously transmitting an increased power level to accommodate fading events on the link, the system provides closed-loop feedback to increase the power level only when fading events occur. Propagation availability can be greatly increased because the threshold at which a link is completely lost is at a much lower received power level. Analysis of the effect of automatic power control on propagation availability will vary for every application based on the control algorithm, geographic region, and service-level agreement constraints.

4.2 EARTH STATIONS

The ground segment of a satellite network is referred to as an earth station or teleport. Earth stations implement the infrastructure required for transmission and reception of signals from the satellite spacecraft. Satellite earth stations consist of a common suite of hardware that is required regardless of station size or operating environment. Figure 4.2 shows a generalized block diagram of the satellite earth station equipment complement. The basic earth station complement consists of the following hardware elements.

4.2.1 Baseband Equipment

Baseband transmission equipment converts and modulates the information signal to be sent into a format that is efficient for transmission over the satellite channel. This

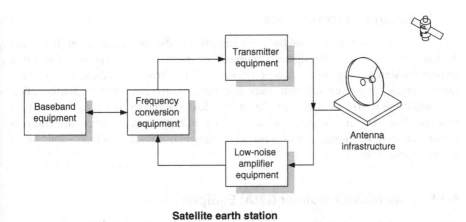

Satellite earth station

Figure 4.2. Generalized satellite earth station equipment complement.

equipment includes (but is not limited to) multiplexing, routing/switching, and modulation hardware. Baseband equipment redundancy configurations are vendor specific and can add significant cost and complexity to network designs.

4.2.2 Frequency Conversion Equipment

The process of signal transmission to the satellite requires baseband signal data to be translated in frequency into the desired frequency band of operation. Baseband signals are "converted" from one frequency to another using frequency converter equipment. Converter equipment can be common to channels carrying traffic ranging from a single satellite transponder to an entire polarization of a satellite. Converter redundancy design is an important consideration for earth station and teleport design.

4.2.3 Transmitter Equipment

Satellite earth station transmitters (also referred to as power amplifiers) amplify the frequency-converted information signal to power levels sufficient for reception and retransmission by the satellite. Transmitter design is a critical item in earth station availability performance. Satellite link design and analysis is a field of study in and of itself and is beyond the scope of this book. Link designers must take into account the required link availability, modulation type, and a host of other design parameters in order to properly select the appropriate transmitter size for a particular earth station. Transmitter redundancy is available in a wide range of configurations. These include nonredundant, one-for-one hot standby, one-for-one cold standby, one-for-N hot standby, one-for-N cold standby, and soft-fail configurations. The specific type of earth station being deployed will dictate the type of transmitter design to be deployed.

4.2.4   Antenna Infrastructure

After the signal to be transmitted has been amplified to the appropriate level (following the link design process), the signal is coupled to a transmission/reception antenna that focuses the signal energy in the direction of the satellite. Antenna infrastructure requires careful engineering since the equipment is exposed to the environment. Earth station antennas range in size from less than 1.0 m for VSAT stations to 20.0 m for hub teleport stations. The size and importance of antenna infrastructure requires attention to foundation design, antenna control, and environmental impacts (such as wind, snow, ice, rain, and others).

4.2.5   Low-Noise Amplifier (LNA) Equipment

Received signals from the satellite spacecraft must be amplified for proper demodulation and recovery. The first stage of receive amplification is the most important in signal recovery. LNAs are used in the first stage of signal amplification because of their excellent noise figure performance. Implementation of LNAs in satellite networks can be done using redundant or nonredundant configurations. Redundant LNA configurations can be challenging to operate and maintain because the LNA must be mounted on the antenna feed in order to achieve the required noise performance. LNA switching systems (required for redundant configurations) can introduce unavailability due to switch failures and design problems that are difficult to predict. LNA equipment is inherently reliable because of its solid-state component complement, and as such, some network operators choose to deploy single-thread LNA subsystems. In those cases, the reliability and availability of the LNA subsystem should be analyzed to ensure that the station performance is consistent with the required performance criteria.

4.3   VSAT EARTH STATIONS

VSAT earth stations comprise a large complement of the currently deployed ground stations in modern satellite networks. The relatively low cost, small size, and high performance of VSAT stations makes them a very cost-effective method for providing communications in remote, difficult-to-serve locations. The low cost and small size of VSAT stations represent an engineering trade-off with respect to availability performance. It is rare to find commercial service VSAT stations operating at frequencies below the Ku-band. As was discussed in the propagation section, operation in the Ku- or Ka-band requires significantly higher fade margins than operation in the C- or X-band. Achievement of the power levels required for very high availability using economical hardware solutions is generally not feasible for VSAT stations. Redundant hardware for the RF or baseband chain for VSAT stations is rare and expensive. Modeling VSAT performance requires consideration of the ground segment (VSAT station), the space segment, and the network topology. This section addresses techniques to model the ground station availability and reliability performance.

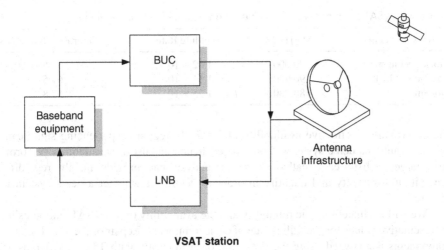

VSAT station

Figure 4.3. Remote VSAT signal chain block diagram.

4.3.1   VSAT Network Equipment

Network equipment utilized in remote VSAT stations most commonly consists of a single-thread RF chain with a relatively small number of baseband modems (usually limited to one or two). Figure 4.3 shows a block diagram of the remote VSAT signal chain.

Availability (and reliability) calculations for the VSAT station signal chain are very straightforward because of their lack of redundancy. The implementation of RF chain components in VSAT network elements combine some RF functions into subsystem blocks. These components are referred to as the block upconverter (BUC) and the low-noise block downconverter (LNB) and are described below.

4.3.1.1   Block Upconverter. The block upconverter combines the frequency upconversion function (translation of baseband signals to RF frequencies for transmission) with the power amplifier function in the RF signal transmission chain. The BUC is a relatively low-cost method to provide two otherwise expensive network functions.

4.3.1.2   Low-Noise Block Downconverter. The low-noise block downconverter combines the frequency downconversion (translation of received RF signals to baseband) with the low-noise amplifier functions. The LNB (like the BUC) is a low-cost implementation of satellite RF hardware.

By combining the frequency conversion subsystems with the power amplification and low-noise amplification stages of the RF chain, the total equipment count is reduced. The availability performance of a VSAT earth station is thus dependent on three major components (BUC, LNB, and modem). VSAT antennas are usually fixed manual mounts and must be erected for resilience to environmental considerations such as wind loading, ice and snow buildup, and corrosion. Properly designed VSAT antennas

Table 4.2. VSAT Component MTTF Values and Computed 5-Year Reliability

VSAT Subsystem	MTTF (h)	Failure Rate	5-Year Reliability (%)
Block upconverter	70,000	1.42×10^{-5}	53.5
Low-noise block	90,000	1.11×10^{-5}	61.5
Modem	80,000	1.25×10^{-5}	57.8

should provide an effective availability of 100%. In regions experiencing significant snowfall, antenna deice systems should be evaluated against the availability performance targets. Historical weather or storm analysis can provide insight regarding precipitation intensity and quantity in order to determine whether a deice system is required.

We will evaluate both the reliability and the availability of the VSAT station using representative values for the failure rate of each component. Repair of the VSAT station components is analyzed using the definition of availability (with TTR modeled as an exponential random variable). This result provides a good approximation of the mean availability performance. Analyses of station availability focused on the variability of the availability performance should use Monte Carlo methods with TTR modeled using an appropriate statistical distribution.

MTTF value for each VSAT subsystem is provided in Table 4.2 . The MTTF values shown are intended to be representative of component performance observed using off-the-shelf hardware but do not reflect specific equipment.

The reliability block diagram for the two-way VSAT station is shown in Figure 4.4 .

Recall that the reliability of a serial combination of components is equivalent to the product of the individual component reliabilities. If we convert the MTTF values provided in Table 4.2 to failure rates and calculate the 5-year component reliabilities, we can use those reliability values to calculate the system reliability for the same time period (recall the definition of reliability: the probability of failure-free performance under stated conditions). For the VSAT system shown, we find

$$R_{\text{VSAT station}} = R_{\text{BUC}} \times R_{\text{LNB}} \times R_{\text{modem}} = 19.0\%$$

Network designers and operators should thus not expect failure-free operation of this particular VSAT system for a 5-year mission life. Eight out of 10 systems would fail in a 5-year period using the components listed. This metric does not necessarily imply that the system must be unavailable, but rather, it simply indicates that the system is not particularly reliable and will require regular maintenance and/or repair. We can simplify

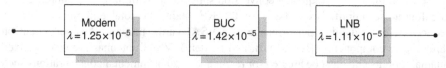

Figure 4.4. VSAT station reliability block diagram.

the block diagram by using the exponential distribution where the total system failure rate is the sum of the individual component failure rates:

$$\lambda_{\text{VSAT station}} = \lambda_{\text{BUC}} + \lambda_{\text{LNB}} + \lambda_{\text{modem}} = 3.79 \times 10^{-5} \text{ failures/h}$$

Having summed the individual failure rates to a single value, we can calculate the availability of the VSAT system by applying the expression

$$A_{\text{VSAT station}} = \frac{\text{MTBF}_{\text{VSAT system}}}{\text{MTBF}_{\text{VSAT system}} + \text{MTTR}_{\text{VSAT system}}}$$

The MTBF of the system is calculated (assuming an exponentially distributed TTF) by applying

$$\text{MTBF}_{\text{VSAT station}} = \frac{1}{\lambda_{\text{VSAT system}}} = 26400 \text{ h}$$

If we assume that the VSAT station failure can be repaired on average in 36 h, the availability of the station is

$$A_{\text{VSAT station}} = 99.86\%$$

This particular VSAT station experiences an average of 11.9 h of outage per year. Although this availability might be sufficient to meet direct-to-home television or even commercial customer demands, the industrial or carrier-grade service requirement may require higher performance than this particular solution can provide.

VSAT station availability can be improved by selecting BUC, LNB, and modem components with lower failure rates or by implementing redundancy. Although the selection of lower failure rate components will improve availability performance, the TTR should also be examined. If we improve the MTTR of the system and assume that a repair can be completed in 18 h (instead of 36 h), the availability is increased:

$$A_{\text{VSAT station}} = \frac{26400}{26400 + 18} = 99.932\%$$

In this case, the annual outage duration is reduced to 6.0 h. It should be clear that without introducing redundancy, the limits of availability performance are quickly reached for the single-thread VSAT station.

4.4 EARTH STATIONS

VSAT earth stations are limited in their ability to achieve very high availability performance by design. A VSAT station is designed to be flexible and inexpensive, achieving good value for locations requiring communications where it is not economical to use other technologies. One way the limited availability performance of a VSAT station can be overcome is to deploy a more traditional earth station. The standard earth station design follows more modular (and scalable) approach to network design.

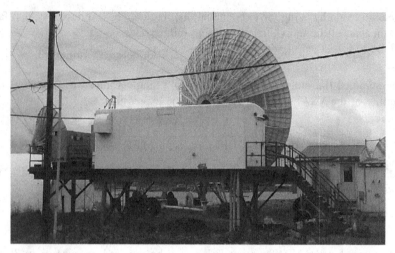

Figure 4.5. C-band satellite earth station constructed in Nome, Alaska.

Frequency conversion, power amplification, low-noise amplification, and baseband modulation are maintained as discrete RF chain components in standard earth station designs. Antennas are generally (but not always) larger in size and have more substantial foundations. Figure 4.5 shows a photograph of a remote earth station constructed in the village of Nome, Alaska.

This earth station operates using one of two C-band antennas and provides a variety of services ranging from traditional voice to high-speed packet data. Dual antennas are implemented in order to ensure restoral capability (across two different spacecraft) in the event of a primary spacecraft failure.

This section presents the technical details associated with improvements possible by deploying a modular, scalable earth station.

Unlike the VSAT station, a standard earth station utilizes modular frequency conversion, power amplification, low-noise amplification, and baseband modulation components. This provides a more scalable growth environment and allows for redundancy to be implemented on each component as required.

4.4.1 Nonredundant Earth Station

Figure 4.6 shows the RF chain block diagram for a typical nonredundant earth station. This earth station is configured to access a single transponder on the spacecraft. As a station needs to access additional transponders, the equipment complement can be increased to accommodate the growth.

Additional frequency converter and baseband equipment is required to access more than one transponder. This design assumes frequency converters operating at 70 MHz intermediate frequency (IF). Other technologies (such as L-band converters) allow access to the entire spacecraft transponder complement but carry performance and operational trade-offs. Both systems utilize the same power amplifier and low-noise

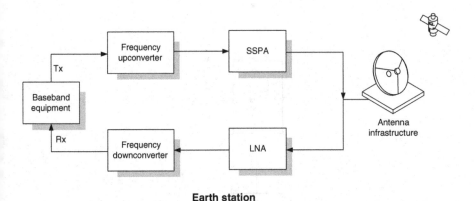

Earth station

Figure 4.6. Typical earth station RF chain block diagram.

amplifier. Redundancy can be implemented on any of the individual frequency converters or modems as is required.

Evaluation of the availability of the nonredundant standard earth station configuration allows us to compare the performance with VSAT systems and follows the same procedure presented in Section 4.1. Table 4.3 shows the failure rates for components shown in Figure 4.6.

Following the same procedure for reliability and availability analysis that was presented in the VSAT station approach, we can develop a reliability block diagram for a nonredundant standard earth station as shown in Figure 4.7.

The total station reliability is the product of the component reliabilities

$$R_{\text{Earth station (NR)}} = R_{\text{U/C}} \times R_{\text{D/C}} \times R_{\text{SSPA}} \times R_{\text{LNA}} \times R_{\text{modem}}$$

where U/C and D/C indicate up- and downconverter, respectively, SSPA indicates solid-state power amplifier, and LNA indicates low-noise amplifier. $R_{\text{Earth station (NR)}}$ designates a nonredundant (NR) earth station. With the given component failure rates, the total 5-year system reliability of the nonredundant system is

$$R_{\text{Earth station (NR)}} = 0.631 \times 0.631 \times 0.747 \times 0.823 \times 0.723 = 17.7\%$$

Table 4.3. Nonredundant Standard Earth Station Component Failure Rates

Earth Station Component	MTTF (h)	Failure Rate (Failures/h)	5-Year Reliability (%)
Solid-state power amplifier	150,000	6.67×10^{-6}	74.7
Low-noise amplifier	225,000	4.44×10^{-6}	82.3
Frequency upconverter	95,000	1.05×10^{-5}	63.1
Frequency downconverter	95,000	1.05×10^{-5}	63.1
Modem	135,000	7.4×10^{-6}	72.3

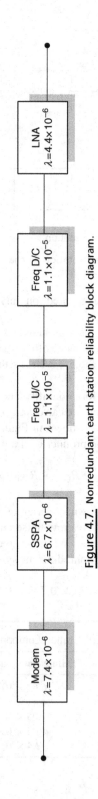

Figure 4.7. Nonredundant earth station reliability block diagram.

The discrete nature of the individual components offers some reduction in the system reliability over the VSAT case (fewer components are present to experience failures and integration of components improves overall reliability). Calculation of system availability follows the same procedure as the VSAT case. We sum the individual failure rates to compute an aggregate failure rate. The total system availability is then computed using the assumed MTTR (36 h) and the aggregate failure rate of the system.

$$A_{\text{Earth station (NR)}} = \frac{\text{MTBF}_{\text{Earth station}}}{\text{MTBF}_{\text{Earth station}} + \text{MTTR}_{\text{Earth station}}} = 99.86\%$$

4.4.2 Fully Redundant Earth Station

Consider an earth station where the importance or revenue associated with the traffic being carried justifies redundancy in all of the network components. Figure 4.8 shows the fully redundant earth station system block diagram.

In this case, the system implements the following redundancies.

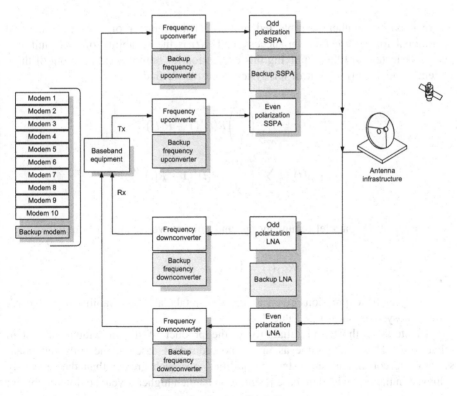

Fully redundant earth station

Figure 4.8. Fully redundant earth station system block diagram.

- One-for-two (1:2) SSPA redundancy
- One-for-one (1:1) upconverter and downconverter redundancy
- One-for-two (1:2) LNA redundancy
- One-for-ten (1:10) modem redundancy

This design assumes that the station access two polarizations of the satellite (horizontal plane and vertical plane). The SSPA and LNA subsystems are 1:2 redundant to reflect this operation. The modem subsystem operates using a protection switching system that houses a single "backup" modem protecting one of the ten operation modems. The modems in this model are assumed to have the same failure rates as those presented in the VSAT section. For our model, we will analyze the worst case in which all ports of the 1:10 modem switching system are populated with modems. Calculation of the reliability of a 1:2 or 1:10 protection scheme is presented in Section 1.3.

$$Pr(S(X) \geq k) = \sum_{y=k}^{n} \binom{n}{y} R(t)^{y}(1 - R(t))^{n-y}$$

where n is the number of protected units, k is the number of units that must be operational for system mission success, and $R(t)$ is the reliability of each unit at a particular instant in time. Applying this expression for both the 1:2 (two out of three systems) and 1:10 (nine out of ten systems) cases, we find

$$R_{1:2}(t) = \sum_{y=2}^{3} \binom{3}{y} R(t)^{y}(1 - R(t))^{3-y}$$

$$R_{1:10}(t) = \sum_{y=9}^{10} \binom{10}{y} R(t)^{y}(1 - R(t))^{10-y}$$

Calculation of the parallel redundant up- and downconverter units is simpler:

$$R_{1:1}(t) = 1 - (1 - R(t))^{2}$$

Applying the expressions given above, we can tabulate the reliability of each earth station subsystem (shown in Table 4.4).

The total earth station reliability is the product of the individual subsystem reliabilities. This is the same as in the nonredundant case. In the fully redundant system, the constituent subsystem reliabilities are much greater than the reliability achieved in the nonredundant case resulting in a much higher 5-year station reliability:

$$R_{\text{Earth station (FR)}} = 0.840 \times 0.994 \times 0.864 \times 0.864 \times 0.237 = 14.8\%$$

Table 4.4. Fully Redundant Earth Station Subsystem Reliabilities

Earth Station Subsystem	Redundancy	5-Year Reliability (%)
Solid-state power amplifier	1:2	84.0
Low-noise amplifier	1:2	99.4
Frequency upconverter	1:1	86.4
Frequency downconverter	1:1	86.4
Modem	1:10	23.7
Modem	1:1	92.3

where $R_{\text{Earth station (FR)}}$ designates a fully redundant (FR) earth station. Thus, the likelihood of a service-affecting failure of the system (here where a system failure is defined as any of the services provided by all 10 modems fails to survive) within 5 years of operation is greatly reduced for each subsystem with the exception of the modem system. Since there are an increased number of modems in service, the likelihood of a failure occurrence is increased. This does not necessarily imply a reduction in availability. Rather it means that the system will require repair in the 5-year life. If the modem system were instead replaced with a 1:1 system, the reliability improves to 57.5% for the same 5-year period, a dramatic improvement over the single-thread earth station design. Although the reliability metric is interesting, it is unlikely that a station would be operated (except for in small VSAT deployments) without preventative maintenance and repair actions being taken to ensure performance continuity. In cases where the system will be repaired promptly, it is also important to analyze the availability of the system. The VSAT and nonredundant earth station availability computations are straightforward and easy to derive. Adding redundant subsystems to the earth station design increases the difficulty of availability computation.

We will calculate the redundant earth station availability in a manner similar to the reliability approach. That is, we will calculate the individual subsystem availabilities and then take the product of those availabilities to determine the overall station availability. Calculation of the redundant subsystem availabilities will utilize the Markov chain approach in our analysis for the 1:1 and 1:2 protected systems. Analysis of the 1:10 modem system will be computed first using the Markov chain approach and second using the Monte Carlo analysis approach. We will assume that the MTTR is 36 h for all station subsystems. The 1:1 up- and downconverter availabilities are the easiest to compute (as derived in Section 1.3 using MTBF = 95,000 h)

$$A_{1:1 \text{ converter}} = \frac{\mu^2 + 3\lambda\mu}{(\lambda + \mu)(2\lambda + \mu)} = 99.99997\%$$

where μ is the repair rate and λ is the failure rate of the up- or downconverter.

The 1:2 SSPA and LNA subsystem availabilities are calculated by applying the five-state Markov model shown in Figure 4.9.

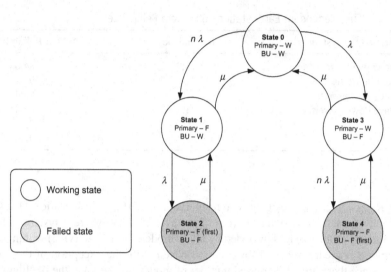

Figure 4.9. One-for-two redundant Markov failure state transition diagram.

This model is developed by first identifying all possible relevant system operational states. Within this model, we identify the following states.

State 0. No failures have occurred.
State 1. One of the two primary units has failed but the backup unit is still working.
State 2. The backup unit has failed in addition to the failure of the primary unit.
State 3. The backup unit has failed while the n primary units continue to function.
State 4. One of the two backup units fails while the backup unit is in a failed state.

Following the analysis technique presented in Section 1.3, we can calculate the solution for the 1:2 redundant system availability. If we write the state transition equations by equating the rates into and out of each state, we resolve the following set of simultaneous equations:

$$\text{State 0}: \quad P_0(n+1)\lambda = (P_1 + P_3)\mu$$
$$\text{State 1}: \quad P_1(\mu + \lambda) = P_0(n+1)\lambda + P_2\mu$$
$$\text{State 2}: \quad P_2\mu = P_1\lambda$$
$$\text{State 3}: \quad P_3(\mu + n\lambda) = P_0\lambda + P_4\mu$$
$$\text{State 4}: \quad P_4\mu = P_4(n\lambda)$$

The last relevant equation that ensures a nontrivial solution is

$$P_0 + P_1 + P_2 + P_3 + P_4 = 1$$

The transition matrix is thus

$$
\mathbb{A} = \begin{bmatrix}
(n+1)\lambda & -\mu & 0 & -\mu & 0 \\
-(n+1)\lambda & (\mu+\lambda) & -\mu & 0 & 0 \\
0 & -\lambda & \mu & 0 & 0 \\
-\lambda & 0 & 0 & (\mu+n\lambda) & -\mu \\
0 & 0 & 0 & -n\lambda & \mu
\end{bmatrix}
$$

The solution of this matrix expression is calculated by applying

$$
\mathbb{P} = \mathbb{A}^{-1}\dot{\mathbb{P}}
$$

where $\mathbb{P}$ and $\dot{\mathbb{P}}$ column vectors represent the state occupation probability and the rate of change of the state occupation probability, respectively. Recall that for the steady-state solution, the rate of change is equal to zero for state transitions. Plugging in the previously defined values for MTBF and MTTR for the SSPA and LNA devices, we can calculate the respective availabilities.

$$
A_{1:2\,\text{LNA}} = 99.999992\%
$$
$$
A_{1:2\,\text{SSPA}} = 99.99998\%
$$

Analysis of the 1:10 protected modem system availability utilizes the same model presented above for the 1:2 protection scheme but replaces the $n = 2$ with $n = 10$. This results in a modem availability of (assuming the individual modem MTBF $= 135,000\,$h):

$$
A_{1:10\,\text{modem}} = 99.99992\%
$$

Table 4.5 shows the availability performance results from each of the individual subsystem analyses.

The aggregate earth station availability is calculated as the product of the individual subsystem availabilities (assuming that the subsystems are statistically independent processes).

$$
A_{\text{Earth station (FR)}} = 99.9998\%
$$

This availability results in an average annual outage of 53 s/year. As discussed earlier, this average value is not representative of a real outage but rather a probability of an event. In the event of a real outage, the duration would more likely be close to the MTTR value of 36 h. Monte Carlo simulation can often impart more insight into the frequency

Table 4.5. Availability Performance for Fully Redundant Earth Station, 36-h MTTR

Earth Station Subsystem	Redundancy	Availability (%)
Solid-state power amplifier	1:2	99.99998
Low-noise amplifier	1:2	99.999992
Frequency upconverter	1:1	99.99997
Frequency downconverter	1:1	99.99997
Modem	1:10	99.99992

of outage occurrences and their durations. The result of this analysis confirms that our fully redundant earth station design achieves the desired highly available performance.

Perhaps most importantly, the results show that the desired availability targets can be achieved using reasonable TTR values. Because most earth stations are deployed in remote, difficult to access areas, the TTR is inherently large. In the analysis presented, 36 h was assumed for the MTTR. This is an aggressive estimate for TTR in remote locations (recall that the TTR includes the time to identify, dispatch, and resolve the service-affecting problem). In many cases, air charters are required in order to access remote earth station problems resulting in long TTR durations. Without implementing redundancy, the limits of achievable availability performance for remote earth stations are easy to identify.

4.4.3 Modular Power Amplifier Systems

Recent developments in power amplifier technology have introduced a third category of redundancy to satellite power amplifier protection schemes. The modular PA system utilizes a number of plug-in power amplifier modules that are phase combined to achieve a desired total output power. Figure 4.10 shows a schematic of the system described.

The system can be configured to operate in one of two possible "modes." Both modes of operation provide a "soft-fail" functionality. Following the failure of a power amplifier module, the system operates with a lowered total power output instead of a complete loss of transmitter output. The first mode operates the power amplifier up to its full rated capacity. In this case, a failure of one of the power amplifier modules results in a complete amplifier system failure (if the system is operated at capacity). This failure occurs because the reduction in total available output capacity drives the power amplifier into saturation, affecting all carriers in the power amplifier. In the second mode, the amplifier is operated up to (but not beyond) a specific, computed power level that allows one or more power amplifier modules to fail without driving the system into saturation. This system can thus be designed to operate with similar availability

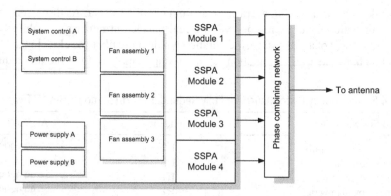

Figure 4.10. Modular satellite power amplifier system block diagram.

performance to the traditional 1:1 redundant system. The risk in operating the system with this type of failure mechanism lies in the fact that the system must not be operating beyond the power reduction factor introduced by the failure of a single power amplifier module (or more than one module in the case of multiple redundancy designs).

For example, consider a modular power amplifier system that operates using four discrete power amplifier modules. The system is designed to operate using three-out-of-four redundancy. If each module provides 100 W of output capacity then the total output capacity is 400 W (4 × 100 W). If we require that the system is capable of supporting operation under the condition that one of the four modules has failed, the total *available* capacity of the power amplifier system is reduced to 300 W. The link must be designed to effectively operate with 1.25 dB reduction in transmit power or alternatively suffer a fade margin loss of 1.25 dB. In addition to this reduction, it is important to be aware of the specifications provided by the equipment manufacturer. The 300 W rating may or may not be the $P_{1\,dB}$ compression value. Furthermore, if the power amplifier is to be operated in a multicarrier environment (with 5–6 dB of output back off), the total available power amplifier capacity is 75 W. The flexibility of configurations allows for multiple different availability targets to be achieved using the same amplifier platform. The availability of both a three-out-of-four and a seven-out-of-eight redundant system is calculated (using Monte Carlo simulation) in Table 4.6. For the purpose of the Monte Carlo simulation, an SSPA module availability of 100,000 h is assumed. The availability of the system is assumed to be completely reliant on the SSPA modules (fan, control, and power system are assumed to have much higher, and thus irrelevant, availabilities). The MTTR of the system is modeled as shown in Figure 4.11.

The results of the Monte Carlo simulation for the two analyses are shown in Figures 4.12 and 4.13. The simulation models each SSPA system life cycle for 20 years and produces a total of 1500 samples.

The y-axis of both figures has been truncated so that the failure occurrences can be better observed. Note that there are a very small number of occurrences (less than 1%) that result in a system availability of less than 100% for the three-out-of-four system configuration. This implies that in most system deployments, the user will not observe a service-affecting outage due to dual module failure.

In the seven-out-of-eight case, the increase in unit count increases the number of experienced failures in the system life cycle. The number of failure occurrences increases to approximately 3% of life cycles. The achieved availability for each system is tabulated in Table 4.6.

When utilizing power amplifiers (or any equipment) that rely upon modular designs sharing a common chassis or other components, it is very important to ensure that the common elements of the device will not negatively affect the overall device availability

Table 4.6. Modular SSPA Simulation Availability Results

Modular SSPA Redundancy	Mean Availability (%)	STD of Availability (%)
Three-out-of-four operation	99.9999	0.001
Seven-out-of-eight operation	99.9997	0.002

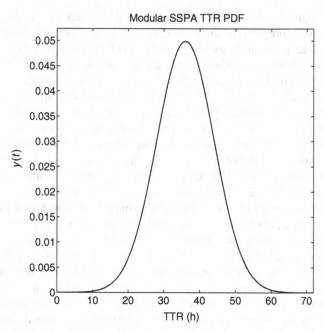

Figure 4.11. Modular SSPA MTTR distribution model.

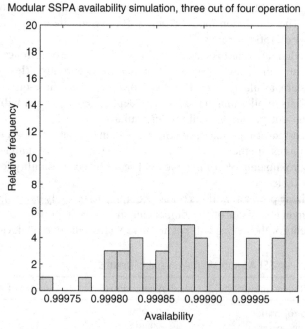

Figure 4.12. Modular SSPA system availability for three-out-of-four configuration.

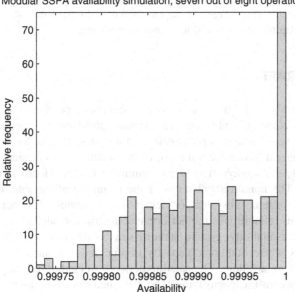

Figure 4.13. Modular SSPA system availability for seven-out-of-eight configuration.

performance. Items such as cooling fans, controller/programming modules, and alarm interfaces should not cause system failures, particularly if they are implemented in a nonredundant manner. Replacement of nonredundant/nonservice-affecting items must be possible while the unit is operational to ensure that availability performance achieves the desired level.

4.4.4 Carrier Management and Monitoring Systems

A class of tools exists in the satellite network market for the management and monitoring of carriers. These tools expand the satellite operator and satellite customer's ability to both operate and maintain their networks at peak performance levels. Although these tools do not provide direct improvements in availability or reliability performance, they provide the user with the necessary feedback to proactively manage network performance. Identification of service impairments within large satellite networks is a challenging task. The presence of underlying interference with a transponder or a carrier is difficult to identify without continuous monitoring of carrier performance metrics (such as E_b/N_0, E_s/N_0, or C/N). Proper implementation (including calibration, maintenance, and continuous improvement) of a carrier monitoring system can allow for alarm thresholding based on carrier performance metrics and interference identification. These tools allow the network operator to more quickly dispatch the proper resources in the case of a service impairment. This in turn may reduce the service degradation from a complete outage to a service impairment. Operators of large satellite networks should consider the use of a carrier monitoring system of some type (COTS or custom) in order to ensure that the performance of the system is consistent with the design targets at all times. Long-term degradation or reductions in system performance can result in complacency in satellite network operation.

4.5 SPACECRAFT

Spacecraft reliability performance constitutes a significant portion of satellite network availability. Spacecraft reliability calculations and considerations can have significant impacts on the overall achieved performance of a system design. Spacecraft must be treated in a manner different from the rest of the satellite network elements. Within a satellite network, the spacecraft (and its communications payload) is the only non-terrestrial item. The nonterrestrial status of the communications satellite means that repair of a failed system, subsystem, or element is not possible. The lack of spacecraft repair necessitates reliability as the performance metric for satellites.

To enhance availability, ground stations can be repointed to a backup/restoral satellite in cases where additional capacity is procured or contracted on a separate spacecraft following a failure. The satellite network (including primary, backup, and any tertiary spacecraft capacity) must be designed for fault tolerance including transponder failure or satellite failure in order to ensure that the network remains available following a failure event. The fault tolerance referenced above is not meant to imply uninterrupted service but rather that the service can be restored following a failure event. It is rare that a catastrophic satellite failure event would result in uninterrupted service because of the low probability of occurrence, high complexity, and high cost of implementation.

It is a common practice for satellite operators to protect a number of active payload spacecraft with an "in-orbit spare" satellite. This spare satellite protects one or more active spacecraft from a complete and catastrophic failure. Under circumstances where the primary communications satellite is no longer capable of fulfilling its mission, either the spare satellite is moved into the primary satellite's position or the ground stations are repointed to the orbital position of the spare satellite (depending on the satellite operator's discretion). While not in operation, the spare satellite might occupy an empty orbital slot or might be coincident with another spacecraft (in which case the payload would be disabled). Figure 4.14 shows a sketch of an arc containing five active satellites and a single in-orbit spare satellite that occupies an empty orbital slot position. The satellites are assumed to occur at 2° longitudinal increments.

The two restoral scenarios discussed above are shown below. In Figure 4.15, the satellite network capacity is restored as the in-orbit spare satellite is moved from its nominal position to the position of the failed satellite.

Figure 4.16 shows the restoral scenario where the spare satellite remains in its current orbital position and the ground stations are repointed to access the new satellite payload.

Both scenarios can be used to restore the satellite network to a fully functioning status. The network topology, satellite operator spacecraft availability, and price dictate which scenario is more appropriate for each user.

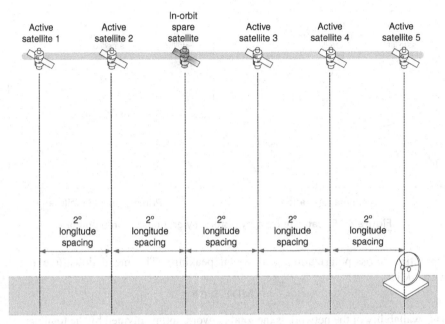

Figure 4.14. In-orbit spare satellite diagram.

Networks with a relatively small number of motorized antennas can consider the use of in-orbit (or other) spare satellites located at different orbital positions (as long as the restoral spacecraft is visible from all ground stations). Restoral of satellite services in this scenario is dependent on how quickly the ground station network operator can command (or manually repoint) the station's antennas to the position of the backup satellite. Consider a network consisting of N-motorized antennas communicating with a primary satellite. Assume that all ground stations are required for the network to be considered to be available. We will calculate the average annual availability of the system if restoral of each station requires an average of 6 h (1 h for problem identification, 3 h for resolution coordination with satellite operator(s), and 2 h for antenna

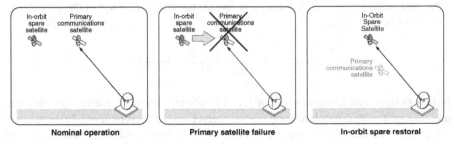

Figure 4.15. Satellite capacity restoral by in-orbit spare move.

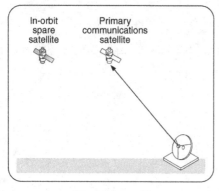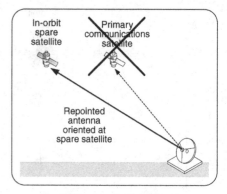

Nominal operation   **Primary satellite failure**

Figure 4.16. Satellite capacity restoral by ground station repointing.

repointing, cross-polarization, and signal peaking). The mean downtime for the network is

$$\text{MDT} = 6\,\text{h}$$

The availability of the network is the total network uptime divided by the total network time:

$$A = \frac{\text{Uptime}}{\text{Total Time}} = \frac{\text{Total Time} - \text{MDT}}{\text{Total Time}}$$

where the "Total Time" is given by

$$\text{Total Time} = 8760\,\text{h}$$

Thus, the network availability *for the year in which the satellite fails* (excluding all ground station failures) is

$$A_{\text{annual}} = \frac{8760 - 6}{8760} = 99.93\%$$

If we assume that only a single satellite failure event occurs in the life of the satellite, we can calculate the satellite capacity availability (assume a 20-year satellite life).

$$A_{\text{life cycle}} = \frac{8760 \times 20 - 6}{8760 \times 20} = 99.997\%$$

Networks operating large numbers of ground stations without motorized antennas can present logistically difficult restoral where antenna repointing is required. In these cases, the restoral plan may be to move the backup satellite to the same orbital position as the failed spacecraft (following a plan previously arranged with the satellite operator). In this case, ground stations remain pointed to the same location while the new satellite is placed into the proper position. The outage associated with this scenario depends on the amount of time required for the satellite operator to command the satellite to its new position in space. The operation to move an in-orbit spare spacecraft can take days or weeks to accomplish (depending on the longitude of the in-orbit spare and the orbital location of the failed satellite) in order to minimize the consumption of the spacecraft's onboard station-keeping fuel. Consider a system where a spare satellite moves into an orbital position $4°$ of longitude from the active satellite's orbital position. Assume that the satellite operator can command the spare satellite to a new position at a rate of $0.75°$/day. The availability of this system is calculated in the same manner as above by first computing the mean downtime and then calculating the availability:

$$\text{MDT} = \frac{4°}{0.75°/\text{day}} \cdot \frac{24\,\text{h}}{1\,\text{day}} = 128\,\text{h}$$

$$A_{\text{annual}} = \frac{8760 - 128}{8760} = 98.54\%$$

The life-cycle availability (also assuming a 20-year life) is

$$A_{\text{life cycle}} = \frac{8760 \times 20 - 128}{8760 \times 20} = 99.927\%$$

The first scenario provides an order of magnitude increase in system availability performance under the stated conditions and is clearly the more desirable network configuration from an availability performance standpoint. The increased cost of antenna motorization and required on-site staffing (particularly for large networks) can often increase the restoral period from the stated 6 h MTTR to a much larger value that might approach the drift time for the in-orbit spare satellite. Analysis and comparison of the two scenarios is an important step in the design of all satellite networks as backup capacity options are considered.

Access to in-orbit spare satellite protection comes at a premium price and can be viewed as a sort of insurance policy against failure. The price paid when services are available can seem excessive and unnecessary but the lack of restoral capacity when a satellite failure occurs can be devastating to a network operator's business. Satellite capacity protection ensures continuity of service for the customer and business for the provider in the case of a satellite failure. In addition to complete satellite failure protection, satellites often employ one or more "reserve" transponders. These reserve transponders are utilized in the case of a transponder failure within the satellite payload. These transponders are considered to be preemptible in the case of a failure of either a spacecraft or a satellite. Access to preemptible capacity is sold at significantly reduced costs to customers because of the lower reliability of the services residing on those transponders.

Spacecraft vendors are often slow to disclose predicted or empirical satellite reliability figures. In cases where this data is available, it can be instructive to assess the risks associated with procuring capacity on satellites with one or more preemptible transponders. Consider a system operating on a primary in-orbit spare satellite with backup capacity on a different in-orbit spare satellite (in the case of a failure). Assume that the mission life (20 years) reliability of each satellite is identical and is 99.9%. The preemptible satellite carrying the primary capacity protects six in-orbit active payload satellites. The probability that the in-orbit spare satellite operates without failure and without being called into service as a spare is given by the system reliability:

$$R_{\text{system}} = R_{\text{SAT1}} \times R_{\text{SAT2}} \times R_{\text{SAT3}} \times R_{\text{SAT4}} \times R_{\text{SAT5}} \times R_{\text{SAT6}} \times R_{\text{spare SAT}} = 99.3\%$$

Although this reliability represents relatively safe gambling odds, in 7-out-of-1000 trials, the system will fail. For noncritical network traffic, this reliability may be sufficient but in cases of business critical data, this value represents an unacceptable risk and procuring backup capacity is a judicious decision. Here, the assumption is that the spacecraft operator will provide the backup spacecraft to restore all services on the failed spacecraft, and that the earth station operator, in his procurement of primary space segment services, makes the decision that such a spacecraft operator's restoration plan is sufficient for his business needs. Based on the calculation above, the probability that a second in-orbit spare (protecting a different complement of active payload satellites) is relatively low (7-out-of-1000 trials). Thus, the decision to procure backup capacity on a second preemptible payload satellite is a reasonable and cost-effective approach to ensuring continuity of service.

Contract negotiations involving capacity protection and restoral are a complex and detailed portion of satellite operations and use. Careful attention must be paid by both the operator and the user of the satellite resources in order to ensure that expectations are met and are consistent with the operators understanding of the contract. The high cost of satellite capacity and protection necessitates expert knowledge of industry practices, reliability analysis, and network availability.

4.6 SATELLITE NETWORK TOPOLOGIES

The discussion of satellite communications so far has been limited to discrete subsystems of satellite networks (earth stations and satellites). This section discusses the calculation of availability and reliability targets for networks of earth stations and satellites. Within the field of satellite communications, variations of two basic network topologies exist. These topologies are

- hub/remote networks and
- point-to-point networks.

The implementation of each of these network topologies might use standardized protocols, proprietary protocols, or a combination of both. Each network type has important considerations for each station type and satellite backup model.

4.6.1 Hub/Remote Networks

One of the most significant and important classes of satellite network in operation today can be generically referred to as the hub/remote network. In this network topology, a hub station (typically utilizing a large-aperture antenna) transmits a large, shared carrier to a number of smaller remote stations. The remote stations are often (but not always) implemented as VSATs. Hub/remote networks are ideal for applications requiring asymmetric capacity such as Internet traffic. Several commercial off-the-shelf (COTS) solutions exist for deploying this network topology within a satellite environment. Figure 4.17 shows a graphical depiction of the hub/remote satellite network topology.

Two different availability results for the hub/remote network topology are presented. In the first scenario, a VSAT network is analyzed that consists of a large-aperture Ku-band VSAT hub station and a number of remote VSAT stations covering a large geographic area with varying antenna aperture sizes (as required by the capacity specifications for each station). Figure 4.18 shows an end-to-end block diagram of this network topology.

In order to calculate the availability (or reliability) of this network, we must consider three major network components. These components are

- hub earth station
- communications satellite
- remote VSAT station

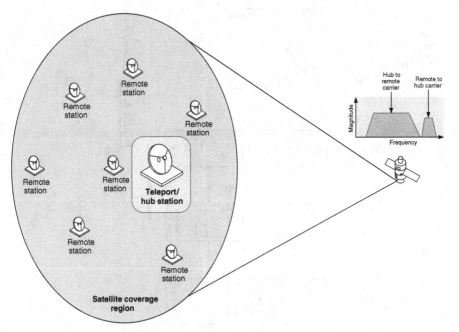

Figure 4.17. Hub/remote satellite network topology.

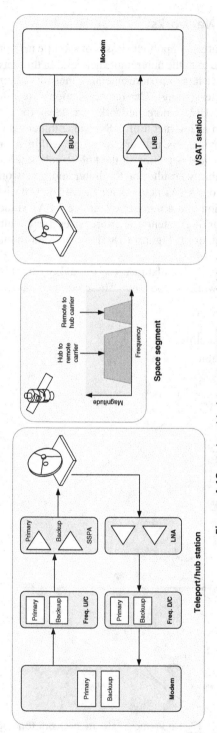

Figure 4.18. Ku-band hub/remote VSAT network block diagram.

162

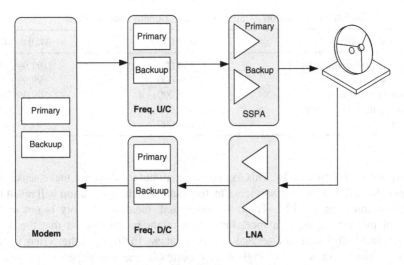

Teleport/hub station

Figure 4.19. Ku-band VSAT hub station block diagram.

For obvious reasons (due to the shared nature of the downstream capacity pool), the hub station must be designed to provide a high availability. Failure of the hub-transmitting station will result in a communication failure at all remote VSAT stations. In addition, the availability of backup capacity on the communications satellite must also be considered as part of the hub station availability. The VSAT station will represent the lowest availability element in this network design.

4.6.1.1 Hub Station Availability. The Ku-band hub station is implemented here as a fully redundant earth station with SSPA, frequency converter, LNA, and modem redundancy. This network design ensures that the availability of the hub station is sufficient to allow the remote station availability to be the only major contribution of unavailability within the network. Figure 4.19 shows a block diagram of this hub station.

Analysis of the hub station availability is performed exactly in the same manner as discussed in the standard earth station section. For simplicity of analysis, the entire hub station baseband equipment complement has been compressed into a single block designated as baseband equipment. This block includes all system control, modulation, and monitoring functions necessary for the system to operate. Full 1:1 redundancy of the baseband equipment is assumed. The availability of this hub station is given by the serial combination of each redundant RF element. Assuming that each RF element is 1:1 redundant, we will use the Markov chain analysis technique to calculate the availability of each subsystem. The availability of each RF element subsystem is shown in Table 4.7. This analysis assumes that the MTTR for all the hub station components is 12 h.

The serial product of the individual RF component availabilities results in a station availability of 99.99999%. This calculated availability results in average annual hub

Table 4.7. Ku-Band Hub Station RF Subsystem Component Availabilities

Earth Station Subsystem	MTBF (h)	Availability (%)
Solid-state power amplifier	150,000	99.9999987
Low-noise amplifier	225,000	99.9999994
Frequency upconverter	95,000	99.9999968
Frequency downconverter	95,000	99.9999968
Modem	135,000	99.9999984

station outage duration of 3 s. This extremely small outage does not make sense in the context of a real system. Any service-affecting outage of the hub station will result in an outage of (on average) 12 h. Thus, the calculated, mean availability is not really a practical measure of performance. Instead, it might be considered that the system operates (annually) without a service-affecting outage. In this case, the system is 100% available. Now assume that a single outage occurs in one annual period and that this outage is 12 h in duration. The availability of the system in this case is thus

$$A_{\text{hub annually}} = \frac{\text{Uptime}}{\text{Total Time}} = \frac{(8760 - 12)}{8760} = 99.86\%$$

If we examine the hub station on availability for a 20-year life cycle and assume that the system experiences (arbitrarily) five outages in the total life of the system, we find that the availability of the hub station for the entire life cycle is

$$A_{\text{hub life cycle}} = \frac{\text{Uptime}}{\text{Total Time}} = \frac{(8760 \times 20 - 12 \times 5)}{8760 \times 20} = 99.97\%$$

This life-cycle availability represents a much more practical target for the hub station. In practice, a properly maintained fully redundant station should rarely fail and might approach the theoretical availability target.

4.6.1.2 Communications Satellite Availability. Given the extremely high theoretical availability calculated for the hub station, the communications satellite must be considered to ensure continuity of service in the case of a spacecraft failure. As discussed previously, a number of different capacity protection options exist. In this case, we will assume that the primary satellite capacity is restored by a secondary satellite operating in a different orbital position than the primary satellite. This requires that the individual remote stations be repointed in the case of a primary satellite failure. Assuming that the system being analyzed serves 150 remote VSAT stations and that each station takes (on average) 36 h to repoint, we can calculate two different satellite service availabilities. In the first case, the satellite is 100% available and no satellite failure occurs throughout the life of the satellite. In the second case, a satellite failure has occurred and requires all of the remote VSAT stations to be repointed:

$$A_{\text{satellite life cycle}} = \frac{(8760 \times 20 - 36)}{8760 \times 20} = 99.98\%$$

In either case (no failure or a single satellite failure), the availability target is on the same order of magnitude as the hub station target for the system life cycle as desired.

The decision to not mitigate a satellite failure using backup capacity on a restoral satellite should not be made lightly. The consequences of a failure can be catastrophic to the business operation and service perception of customers.

4.6.1.3 VSAT Station Availability. The final component of the hub/remote network topology is the VSAT station. For this analysis, we will directly apply the results calculated in the VSAT section. The availability results calculated in that section is reproduced as

$$A_{\text{VSAT system}} = 99.86\%$$

Recall that the VSAT system assumes an MTTR of 36 h.

4.6.1.4 Network Availability. Using the serial product of the individual network subsystem elements, we can calculate a number of different availability scenarios as in Table 4.8. These life-cycle scenarios are enumerated below.

- No hub station failures, no satellite failures
- Five hub station failures, no satellite failures
- No hub station failures, one satellite failure
- Five hub station failures, one satellite failure

Clearly, a number of other availability scenarios (involving the number of hub station failures) might be envisioned. The results presented here represent the best- and worst-case availability performance to be expected from the system. The remote stations are assumed to follow the theoretical predicted performance targets.

It should be clear that a properly designed hub and satellite network components do not drive the total hub/remote network availability. The remote VSAT station is the most significant factor in this network design. This has two implications. The first implication is that improvements to network availability should be focused on the VSAT station.

Table 4.8. Calculated Hub/Remote Network Availabilities

Scenario	Network Availability (%)
No hub station failures, no satellite failures	99.86
Five hub station failures, no satellite failures	99.83
No hub station failures, one satellite failure	99.84
Five hub station failures, one satellite failure	99.81

The second implication is that hub and satellite network components will not contribute significantly to network unavailability unless a major component failure occurs.

4.6.2 Point-to-Point Networks

Point-to-point networks represent a substantial portion of the total satellite communications networks deployed today. In a point-to-point network topology, communication exists exactly between two ground stations. All three network elements involved (both ground stations and the satellite) exhibit equal importance in network availability performance. Bidirectional point-to-point networks require two discrete signal chains (Stations A–Z and Stations Z–A). Figure 4.20 shows the bidirectional signal path for a sample point-to-point network.

Calculation of network availability for point-to-point networks follows the same technique used in the hub/remote topology. The primary difference lies in the bidirectional symmetry that is typically found in point-to-point networks. Assuming a fully redundant station design identical to the hub station design shown in the hub/remote network, we have two earth stations with life-cycle availabilities equal to

$$A_{\text{ground station life cycle}} = \frac{\text{Uptime}}{\text{Total Time}} = \frac{(8760 \times 20 - 12 \times 5)}{8760 \times 20} = 99.97\%$$

If we also assume that the communications satellite availability is the same as discussed in the hub/remote network design:

$$A_{\text{satellite life cycle}} = \frac{(8760 \times 20 - 36)}{8760 \times 20} = 99.98\%$$

The bidirectional point-to-point network availability is the serial product of the two ground station availabilities and the satellite availability. Assuming a 20-year life cycle, the network availability averaged across the system life is

$$A_{\text{PTP network}} = \left(A_{\text{GS life cycle}}\right)^2 \times A_{\text{satellite life cycle}} = 99.92\%$$

Given the extremely high availability achievable by the fully redundant station, examination of the 20-year reliability of the station can provide performance insight in systems with a large number of deployed stations. The reliability of each earth station subsystem is given in Table 4.9.

The total system reliability is the serial product of the individual subsystem reliabilities:

$$R_{\text{GS life cycle}} = R_{1:1 \text{ SSPA}} \times R_{1:1 \text{ LNA}} \times R_{1:1 \text{ upconv}} \times R_{1:1 \text{ downconv}} \times R_{1:1 \text{ modem}} = 1.5\%$$

Since the probability that the station survives 20 years without a failure (by definition of reliability) is only 1.5%, it is almost certain that every station will require repair or

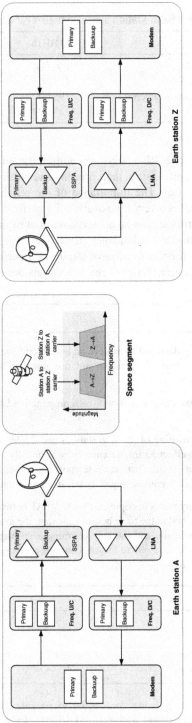

Figure 4.20. Bidirectional point-to-point satellite network block diagram.

Table 4.9. Fully Redundant Earth Station Subsystem 20-Year Reliability

Earth Station Subsystem	MTBF (h)	20-Year Reliability (%)
1:1 solid-state power amplifier	150,000	52.5
1:1 low-noise amplifier	225,000	70.7
1:1 frequency upconverter/downconverter	95,000	29.1
1:1 modem	135,000	47.2

replacement of at least one RF subsystem within the life of the station. Proper operational procedures will ensure that preventative maintenance efforts mitigate outages due to wear and tear. The importance of the reliability result shown is that although the system achieves a very high availability performance, the engineer should not expect that this performance can be achieved without paying careful attention to the MTTR and to ensuring that the redundancy of each subsystem performs as designed. Failures will occur, and in order to achieve the desired performance, repairs must be performed promptly to ensure that performance targets are met.

QUESTIONS

4.1. A satellite link is being designed to achieve an end-to-end availability of 99.99%. Explain why Ku-band and Ka-band frequencies are not practical for achieving this level of propagation availability.

4.2. Describe the conditions under which multipath propagation is a concern for satellite earth station design.

4.3. A satellite network consisting of 75 earth stations and 2 teleport locations experiences (on average) three service affecting interference events annually. The events last (on average) 4 h and typically affect services between a teleport and a remote earth station. What is the annual service availability considering only the impact of interference outages?

4.4. An oil and gas company wants to deploy a 15-site VSAT network with network component MTBF values as described in the table below. Calculate the availability of communications between a VSAT remote site and the hub assuming a propagation availability of 99.8%. What sparing levels are required to ensure continuity of operation on an annual basis? Assume an MTTR of 18 h.

Component	MTBF (h)
BUC	50,000
LNB	75,000
Modem	85,000
Hub (redundant)	215,000

4.5. A C-band earth station is being constructed to provide carrier-class communications between a remote village and a centralized teleport location. If the required station availability is 99.99%, what redundancy level is required for each of the components in the table below to achieve the target performance if the MTTR for the station is 12 h?

Component	MTBF (h)
LNA	115,000
SSPA	85,000
Frequency converter	90,000
Modem	125,000

4.6. What is the service availability for a link between two identical stations (as described in Q4.5) with a designed propagation availability of 99.98%?

4.7. A modular earth station power amplifier system is being considered for deployment in place of an existing 1:1 power amplifier system. Assume that each power amplifier in the 1:1 system has an MTBF of 75,000 h and costs $60,000 per amplifier. Modular power amplifier system modules have an MTBF of 40,000 h and cost $20,000 each. Which system is preferable if the modular system is designed for three out of four operation? Assume an MTTR of 24 h for the SSPA systems.

4.8. A satellite network operator sells space segment capacity with a claimed 15-year reliability of 99.99%. Assuming an exponentially distributed TTF, what is the probability that the satellite will survive for 10 years without a failure?

4.9. If the satellite described in Q4.8 is operated with in-orbit protection (assuming that the replacement satellite is positioned in the same orbital position as the failed satellite with a drift time of 72 h), what is the availability of satellite service? Assume that only one failure occurs in the life of the satellite.

5

MOBILE WIRELESS NETWORKS

Mobile wireless networks make up a significant portion of modern communications system research, infrastructure, and spending. Consumer demand for ever-increasing system performance (improved call quality, network coverage, and data rates) drives investments by manufacturers, operators, and customers. Central-to-system performance is the design of mobile wireless networks with defined availability targets. As more consumers move away from wireline telephone service, the requirement for uninterrupted wireless service is intensified. With millions of subscribers and a large installed base of equipment, network engineers must have a well-developed understanding of system availability, network sparing, and design trade-offs. Mobile wireless networks consist of a number of central network elements that are required for system operation. Availability and reliability analysis of mobile wireless networks follow techniques similar to those presented in the previous chapters. Many subscribers are distributed geographically across many base stations. Backhaul between the base transceiver station (BTS) elements and network core elements requires particular attention. Backhaul provided by off-net (third-party) network operators must have contractual SLA availability and performance metrics defined. Backhaul network designs under the control of wireless network operators must be given the same attention as other network mobile wireless elements or subsystems. Failure to consider

Telecommunications System Reliability Engineering, Theory, and Practice, Mark L. Ayers.
© 2012 by the Institute of Electrical and Electronics Engineers, Inc. Published 2012 by John Wiley & Sons, Inc.

availability as a central network design goal on modern cellular wireless networks can result in poor performance, increased operational costs, and lost revenue.

A variety of different mobile wireless network technologies and standards are deployed throughout the world. In general, the network topology of deployed technologies is very similar. The most significant differences in the technologies lie in the air interface. Availability and reliability considerations remain similar throughout all the mobile wireless networks. Emerging technologies such as long-term evolution (LTE) are evolutionary extensions of existing technologies and, as such, the analysis techniques presented here will be applied with minimal modifications. The established nature of existing technologies allows a comprehensive review of availability, reliability, and operational impacts of large-scale networks.

Although there are variations in the design and implementation between different mobile wireless technologies (such as GSM, CDMA, UMTS, LTE, and others), the fundamental building blocks and design philosophies remain consistent across all technologies. In addition, there is a fair amount of variation in deployments stemming from vendor interpretation of wireless standards. In consideration of these variations, the information presented is intended to provide guidance and direction regarding the analysis and design of reliable mobile wireless networks.

5.1 MOBILE WIRELESS EQUIPMENT

Mobile wireless networks utilize a complex interconnected network of equipment that performs the functions necessary to provide wireless voice and data services to subscribers. The network elements control call switching, data traffic flow, end-to-end call setup, and network management. This section discusses each mobile wireless network element, the redundancy techniques employed, and availability analysis techniques. Figure 5.1 shows a block diagram for a global system for mobile communications (GSM) wireless cellular network. Although the figure is specific to GSM networks, the basic network elements identified are common to GSM, CDMA, UMTS, and other network designs. The basic network switching, cell site backhaul, radio resource, and

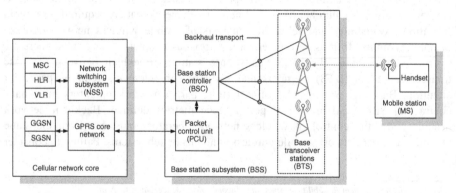

Figure 5.1. GSM network block diagram.

subscriber interface network elements are common to most mobile wireless network types. Vendor-specific implementations of mobile wireless networks contribute significantly to the redundancy options available to network designers.

5.1.1 Network Switching Subsystem (NSS)

The NSS is the subsystem of the mobile wireless network responsible for call switching and mobility management. The NSS consists of a number of discrete network elements. Some of the network elements are integrated into other elements while others are stand-alone devices. The core network elements within the NSS are:

- mobile switching center (MSC)
- home location register (HLR)
- visitor location register (VLR)
- authentication center (AuC)
- equipment identity register (EIR)

Each of these elements represents a network function critical to mobile wireless service delivery. The following section provides a qualitative description of the network elements and the commonly deployed redundant configurations.

5.1.1.1 *Mobile Switching Center (MSC).* MSC equipment is responsible for routing voice and short message system (SMS) traffic, end-to-end call management, and mobility management. The MSC is a core element of the NSS and is fundamental to operation of the mobile wireless network. The MSC is most commonly deployed as a one-for-one redundant network element that serves an entire region. Large carriers often deploy multiple geographically diverse MSCs in order to distribute the network load. This distributed topology improves network availability. The failure of a single MSC in a multiple MSC topology does not lead to a catastrophic network failure. Due to the high cost of MSC network elements, multiple geographically diverse MSC devices are only practical for large subscriber counts.

A small number of specialized network vendors offer MSC equipment capable of small-scale distributed network topologies. Figure 5.2 shows a sketch of the small-scale distributed MSC network topology. We will calculate the MSC availability first for a system utilizing single MSC one-for-one redundancy and a geo-diverse system with 2 one-for-one MSC devices. In this model, the subscriber load is equally split between the two diverse MSCs.

Distributed MSC networks offer improved network availability in areas where local calling must be maintained during backhaul outages. Specifically, networks using satellite backhaul can benefit from a distributed MSC deployment. When the backhaul fails in this scenario, the wireless-to-wireless calls are still possible between handsets served by the isolated MSC as in Figure 5.3.

We will calculate the wireless-to-wireless service availability for callers in a satellite-served community and compare this to the availability of wireless-to-wireless

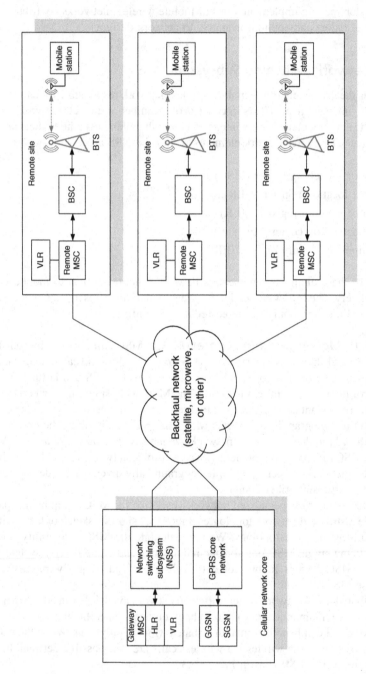

Figure 5.2. Distributed MSC network block diagram.

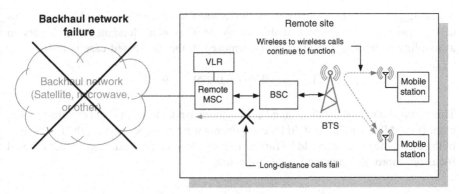

Figure 5.3. Distributed MSC failure scenario and service continuity.

service availability for a centralized MSC topology. In the centralized MSC topology, the availability of the wireless service can be written as

$$A_{service} = A_{core} \times A_{backhaul} \times A_{BSC/BTS}$$

where $A_{service}$ is the availability of the wireless service. Service is assumed to be available if a subscriber can successfully make a call (wireless-to-wireless in the local, satellite-served community) at any moment in time. A_{core} is the availability of the core elements. This includes all NSS functions required for call completion and is assumed to be redundant with an availability of 99.995%. $A_{backhaul}$ is the availability of the backhaul communications channel. In this example, the backhaul is assumed to be provided by a Ku-band satellite circuit with an availability of 99.8%. $A_{BSC/BTS}$ represents the availability of the base station transceiver and its traffic processing functions (base station controller) of the wireless network (the radio access network or RAN). The BSC and BTS equipment implement traffic flow and over-the-air interfaces. In this example, we will assume that the availability of the BSC/BTS equipment is 99.9%, including the availability of the cell site backhaul/transport system linking the base station to the BSC and remote MSC. In this simplified case, we assume only equipment availability of the cell system RAN elements. It does not include the base station accessibility (the probability that the BS has adequate capacity to process the call) or the call retainability (probability that a call once established will be held up for the duration of the call and not be dropped).

The total system availability is thus

$$A_{service} = 0.99995 \times 0.998 \times 0.999 = 99.7\%$$

Implementing the distributed MSC functionality changes the expression for calculating the system availability:

$$A_{service} = A_{Remote\ MSC} \times A_{BSC/BTS}$$

where $A_{\text{Remote MSC}}$ is the availability of the remote (distributed MSC) and $A_{\text{BSC/BTS}}$ is as defined previously. Assume that the remote MSC is nonredundant and achieves an availability of 99.9%. The availability achieved in the distributed case is thus

$$A_{\text{service}} = 0.999 \times 0.999 = 99.8\%$$

This result shows that the distributed case is more available, even if only slightly in this case. If the MSC and the BSC/BTS components were made redundant, the availability of the distributed system could be further improved while the centralized system should focus on improving backhaul availability first.

5.1.2   Base Station Controller (BSC)

The base station controller in cellular networks is responsible for managing the radio resources and handset requests provided by the base transceiver stations with which it is associated. BSCs in large-scale mobile wireless networks can manage hundreds of individual base transceiver stations (BTSs). The criticality of the BSC lends itself to redundancy in its implementation. Most mobile wireless networks implement a hot-standby BSC with one-for-one redundancy. Figure 5.4 shows a sketch of the base station subsystem (BSS) with a single BSC and a number of associated BTS units.

Base station controllers can be configured as stand-alone devices or as devices that also incorporate other functions and the base station controller function. Some BSC devices incorporate the "transcoding" function directly into the BSC while others utilize a separate device to perform the transcoding function. Transcoding refers to the operation of changing the voice signal from one coding scheme to another.

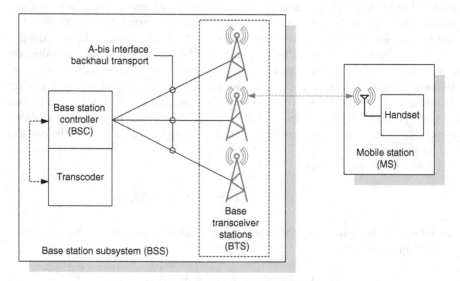

Figure 5.4. Base station subsystem block diagram.

This operation is performed when transferring a voice call from a wireless device (which will use some form of low bit rate vocoder) to a traditional wireline switch or vice versa (where a standard 64 kbps PCM voice waveform encoder is used).

5.1.3   Base Transceiver Station (BTS)

The base transceiver station equipment includes the transceiver radio hardware, antennas, and modulation functionality necessary to implement the over-the-air interface of the cellular network. Implementation of the BTS varies dramatically between wireless technologies. GSM, CDMA, UMTS, and LTE all use significantly different BTS implementations in order to achieve the performance required by each standard. The BTS devices are made up of a controller subsystem and one or more transceivers. Control functions performed by the BTS vary depending on the technology deployed. In CDMA 2000, GSM, and UMTS systems, the BSC or RNC perform most control functions while in LTE a significant number of control functions are performed right in the base station. The transceiver is responsible for transmission and reception of signals over the air. In order to enhance performance and increase capacity, base station designers sometimes "sectorize" the coverage from a particular base station. This sectorization allows for increase signal energy in the direction of the sector and for additional capacity in that sector. BTS devices commonly implement more than one transceiver (TRX) unit per BTS or sector. The quantity of TRXs is dependent upon system capacity requirements but can be limited by the amount and frequency reuse that can be tolerated. BTS hardware is not frequently deployed in a hot-standby configuration. Redundancy is more often obtained by the statistical multiplexing effect obtained as additional TRX units (and ultimately voice channels), which are added to the system. Consider a generic wireless network with three base stations. Each base station has one sector and two channels (comprising perhaps one or more TRXs per channel). The radio hardware for each channel is a slide-in module and is hot swappable (replaceable while the system is in service). Figure 5.5 shows a sketch of this base station network.

The redundancy configuration of this network requires the analysis to consider a number of different failure/restoral scenarios. The analysis is more complex than in previous examples because in the base station case, the equipment is soft-fail redundant. Soft-fail redundancy (in this case) means that if one of the two TRX modules fails, the system does not cease to provide service but rather provides service at an impaired capacity. In order to complete this analysis, we must define the conditions for impairment.

Consider the expression shown below:

$$
\begin{aligned}
Pr(\text{Service}) = \ & Pr(\text{TRX}_1 \text{ and TRX}_2 \text{ working}) \\
& \times Pr(\text{Service if TRX}_1 \text{ and TRX}_2 \text{ working}) \\
& + Pr(\text{One TRX failure}) \\
& \times Pr(\text{Service if TRX}_1 \text{ or TRX}_2 \text{ not working}) \\
& + Pr(\text{Both TRX failed}) \\
& \times Pr(\text{Service if TRX}_1 \text{ and TRX}_2 \text{ failed})
\end{aligned}
$$

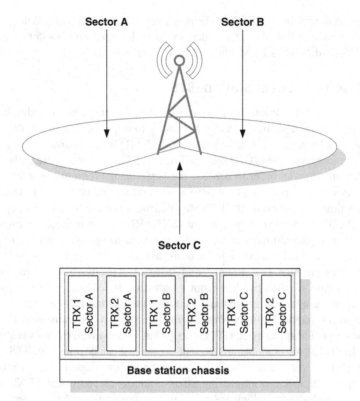

Figure 5.5. Mobile wireless base station TRX configuration.

We will abbreviate the expression above with the following notation:

$P_0 = Pr$(TRX$_1$ and TRX$_2$ working)
$P_1 = Pr$(One TRX failure)
$P_2 = Pr$(Both TRX failed)

Solution to the problem requires resolution of the probability of each of the three system states as well as the probabilities of service availability in each system state. The example presented here is solved by using Markov chain analysis to determine the probability of occupation for each of the three system states. The problem can be simplified to a determination of the system state occupation and the probability that service is available in each of those system states. In our previous discussions of hot-standby availability, the sum of the probability of service in each of the three states was 100%. Due to the impaired system states occurring when one of the TRX modules fails, the probability of service is modified when one or two of the TRX modules fails. Figure 5.6 shows a sketch of the Markov chain state transition diagram for the soft-fail TRX modules. It is assumed that if both modules fail, the resources are available to repair both TRX modules to as good as new condition in the same amount of time that a

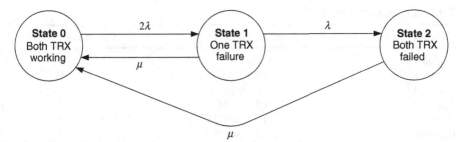

Figure 5.6. Markov chain state transition diagram for BTS TRX modules.

single TRX repair takes. In addition, it is assumed that when a dual TRX module failure occurs, the system is returned to an operational state where both TRX modules are functioning.

Solution of this Markov chain follows the same technique presented in Chapter 2. The solution to the Markov chain analysis is presented below for a system with $\text{MTBF}_{\text{TRX}} = 85,000\,\text{h}$ and $\text{MTTR} = 24\,\text{h}$.

$$P_0 = \frac{\mu}{2\lambda + \mu} = 0.99944$$

$$P_1 = \frac{2\lambda\mu}{(\lambda + \mu)(2\lambda + \mu)} = 5.642 \times 10^{-4}$$

$$P_2 = \frac{2\lambda^2}{(\lambda + \mu)(2\lambda + \mu)} = 1.593 \times 10^{-7}$$

Determination of the service availability during each system state relies upon definition of two service conditions. Coverage is defined in this analysis to mean that the handset in question has sufficient signal power (in both the receive and the transmit directions) to complete a call or data session. Congestion is defined to mean that the primary or tertiary base station associated with the handset has insufficient capacity to serve the handset's requests.

In the P_0 case, the probability of coverage is 100% since the base station providing service is the closest base station and the coverage is assumed. Congestion is not assumed in this case since it is expected that the base station is functioning within the designed call blockage specifications. We can write the P_0 probability expression as

$P_0 = Pr(\text{TRX}_1 \text{ and TRX}_2 \text{ working}) \times Pr(\text{Coverage}) \times Pr(\text{No Congestion})$
$P_0 = 0.99944 \times 1.0 \times 1.0 = 99.944\%$

In the case of the system state where one of the two TRX modules has failed (P_1), the probability expression consists of the state occupation probability multiplied by probability of two possible service deliveries.

$$P_1 = Pr(\text{One TRX failure}) \times Pr(\text{Service})$$

Table 5.1. BTS Regions and Coverage

Coverage Region	Base Station Coverage	Percentage of Area (%)
Region 1	BTS A	23
Region 2	BTS B	23
Region 3	BTS C	23
Region 4	BTS A and C	9
Region 5	BTS B and C	9
Region 6	BTS A and B	9
Region 7	BTS A, B, and C	4

where Pr(Service) is given by

$$Pr(\text{Service}) = Pr(\text{No Congestion on Primary BTS})$$
$$\times \ Pr(\text{Coverage on Primary BTS}) + Pr(\text{Congestion on Primary BTS})$$
$$\times \ Pr(\text{Coverage on Another BTS})$$

The probability of congestion on the primary cell given a single TRX failure is assumed to be 25%. Coverage probability is assumed to be 100% for the primary cell. Probability of coverage from another BTS is a computed value based on the coverage region of each BTS. Figure 5.7 shows a simple sketch of the three base stations. In this diagram, it is assumed that the base stations are positioned equidistant from each other forming an equilateral triangle. The coverage of each base station is further assumed to be identical (as shown). The probability of coverage from any of the three base stations can be calculated for a set of different mobile handset locations. These locations are defined by overlapping coverage regions. A total of seven distinct regions exist in this example. The regions are defined in Table 5.1 with the percentage of the total covered area for each region indicated.

The availability of Regions 1–3 is lower than that of Regions 4–6 while Region 7 has the highest availability of all of the handset locations. Region 7 has two coverage options in the case of a dual TRX failure while Regions 4–6 have one option during a dual TRX failure. The service availability will vary dependent on the location of the handset within the coverage area of the three base stations.

The simplistic approach presented here is intended to indicate a generalized approach for calculation of BTS service availability. Clearly, more sophisticated coverage analysis techniques would be required for real systems. Extension of the simple symmetric, circular coverage approach to a more complex analysis is a simple matter but requires propagation analysis for specific regions.

$$Pr(\text{Service}) = (1.0 - 0.25) \times 1.0 + 0.25 \times Pr(\text{Coverage on Another BTS})$$

The probability of "Coverage on Another BTS" is dependent on the location of the subscriber. If we assume that the subscriber's location within the coverage area is uniformly distributed, we can calculate a probability of coverage using the percentages provided for overlapping BTS coverage regions. Within the symmetric model regions

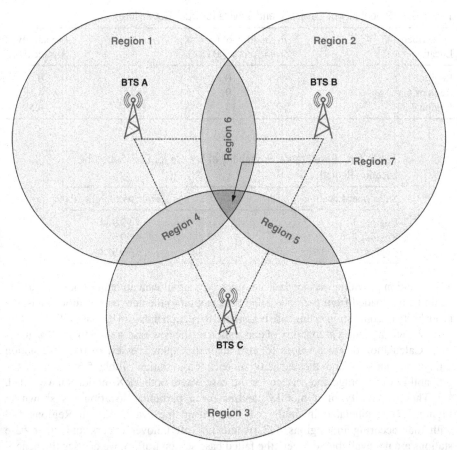

Figure 5.7. Base station overlap and probability of coverage by multiple stations.

1–3, 4–6, and 7 represent different availability performance. The probability of coverage for each of the three conditions and the probability of service for that condition are provided in Table 5.2.

In the case where both TRX modules have failed, the probability that service will be available is dependent on the subscriber's position (and thus the coverage probability) and the probability that the base station providing service is congested. It should be clear that if

Table 5.2. Probability of Coverage and Service for One TRX Failure

Subscriber Location	Probability of Coverage from Another BTS (%)	Probability of Service (%)
Regions 1–3	0	75.0
Regions 4–6	9	77.3
Region 7	13	78.3

Table 5.3. Probability of Coverage and Service for Dual TRX Failures

Subscriber Location	Probability of Coverage from Another BTS (%)	Probability of Service (%)
Regions 1–3	0	0
Regions 4–6	9	4.5
Region 7	13	6.5

Table 5.4. Base Station Probability of Service for Each Subscriber Location Region

Subscriber Location	Service Availability (%)
Regions 1–3	99.944
Regions 4–6	99.987
Region 7	99.988

a base station providing service fails completely, a significant amount of network traffic could be off-loaded from the now-failed base station to the new base station. Thus, the probability of congestion in this case is assumed to be higher than in the single TRX failure case. We will assume a probability of congestion on the new base station to be 50% in this case. Calculation of these values for real analysis requires review of the base station utilization statistics and the capacity of each base station. Table 5.3 provides the probability of coverage and service for the case where both TRX modules have failed.

The availability of a mobile handset for a particular location (as shown in Figure 5.7) is tabulated in Table 5.4. Comparing the availability in Regions 1–3 with that occuring in Regions 4–7 (where no soft failover occurs and other base stations are not available to accept the failed base station traffic), we can see the benefit of the redundancy implementation.

5.2 MOBILE WIRELESS NETWORK SYSTEMS

Mobile wireless networks rely heavily on transport networks for transmission and routing of signals carrying voice and data in the network. These networks interconnect the MSC, BSC, and BTS devices (among others) utilizing time-division multiplexed (TDM) and packet-based (Ethernet) circuits and networks. The transport technology used to interconnect mobile wireless hardware is dependent on manufacturer's specifications and the generation of the hardware. Almost all network hardware is converting to Ethernet or packet-based network communications. Although current and next-generation hardware relies on Ethernet transport, a large installed base of network elements relies upon legacy TDM transport for the delivery of inter-element communications.

Mobile wireless network elements are intrinsically joined with their associated transport networks from an availability standpoint. Mobile wireless networks implemented to achieve high availability will fail if the same considerations are not addressed

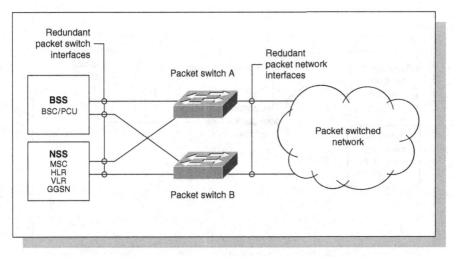

Figure 5.8. Network switching subsystem packet switching redundancy.

for the underlying transport technologies. Figure 5.8 shows a core network infra-structure in which the discrete NSS components are connected to a pair of high-availability packet switching elements. Clearly, the packet switching elements have the same network availability criticality as the MSC and its associated voice and data switching elements. The network elements connecting to the switches have been simplified in this network diagram for clarity.

As discussed in Chapter 2, a system is at most as available as its most unavailable element. Thus, if the system utilizes packet switching and routing elements that are not designed with the same performance targets as the mobile wireless network elements, the system will not achieve the desired performance. Of particular interest is the analysis of systems utilizing backhaul circuits provided by third-party network operators. In these cases, the network operator should be careful to ensure that contractual SLAs achieve the desired availability results.

Consider a stand-alone GSM network with centralized NSS and BSC functionality serving a total of 150 BTS nodes. This network is constructed using a combination of Ethernet over fiber-optic transport backhaul, leased TDM backhaul, and short-haul microwave network backhaul. The target subscriber availability is 99.5%. That is, when a subscriber attempts to make a call using a mobile station, 99.5% of request results in a successful call (5 out of every 1000 calls fail). We are going to consider a simplified model in which a successful call is implied by the NSS and the BSS systems being available at the time of the call. Figure 5.9 shows a sketch of this example wireless network.

The availability of the NSS, the BSS, and underlying packet switching network (backhaul network) must combine to achieve a total minimum system availability of 99.5%. Breaking this particular problem into its most basic elements, we can create a serial system availability expression defined as

$$A_{\text{subscriber}} = A_{\text{NSS}} \times A_{\text{BSC}} \times A_{\text{backhaul}} \times A_{\text{BTS}} = 99.5\%$$

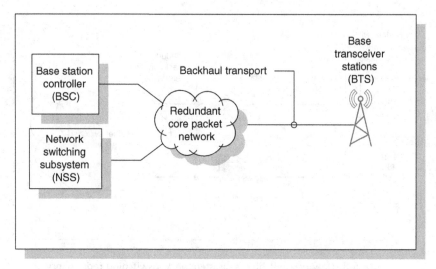

Figure 5.9. Example GSM cellular wireless network.

Examining each individual subsystem is instructive in system analysis. It often provides engineers with insight into system weaknesses and flaws. In this problem, we will assume that the NSS and the BSC (including the packet switching) functions are implemented with redundant components. If we assume that the MTBF of MSC is 90,000 h, the MTBF of the BSC is 115,000 h, and the MTTR for the core elements is 8 h, we can calculate the availability of each one-for-one hot-standby redundant systems to be

$$A_{NSS} = \frac{\mu^2 + 3\lambda_{NSS}\mu}{2\lambda_{NSS}^2 + 3\lambda_{NSS}\mu + \mu^2} = 99.999998\%$$

$$A_{BSC} = \frac{\mu^2 + 3\lambda_{BSC}\mu}{2\lambda_{BSC}^2 + 3\lambda_{BSC}\mu + \mu^2} = 99.9999990\%$$

Clearly, the NSS and the BSC subsystems easily achieve the desired availability target. For the purposes of further analysis, the contribution of these components to system unavailability can be neglected. Assume that the BTS radio devices are implemented in a single-thread configuration with one or more (up to three) sectorized TRX modules. The worst-case availability will be achieved in the case of a BTS with three TRX modules (assuming that this BTS has no overlapping coverage with adjacent cells). The BTS has an MTBF of 125,000 h while each TRX module has an MTBF of 75,000 h. The MTTR of the BTS is necessarily longer (due to the geographic distribution of cell sites) at 36 h. Thus, the BTS availability can be determined by applying

$$A_{BTS\ Site} = A_{BTS} \times A_{TRX}^3$$

where

$$A_{BTS} = \frac{MTBF_{BTS}}{MTBF_{BTS} + MTTR_{BTS}} = 99.97\%$$

$$A_{TRX} = \frac{MTBF_{TRX}}{MTBF_{TRX} + MTTR_{BTS}} = 99.95\%$$

Having calculated the availability of the individual units, we can calculate the total BTS site availability:

$$A_{BTS\ Site} = A_{BTS} \times A_{TRX}^3 = 99.83\%$$

The target availability of 99.5% is still within our grasp since the NSS and the BSC systems are an order of magnitude more available than the nonredundant BTS subsystem. Achievement of the target availability is now dependent on specification of the backhaul availability:

$$A_{subscriber} = A_{NSS} \times A_{BSC} \times A_{backhaul} \times A_{BTS} = 99.5\%$$

$$A_{backhaul} = \frac{0.995}{A_{NSS} \times A_{BSC} \times A_{BTS}} = \frac{0.995}{0.9983} = 99.67\%$$

Using the analysis presented in this example, wireless network engineers can specify backhaul design requirements or SLA requirements (in the case of leased backhaul) to other departments to ensure that the total system availability is achieved.

QUESTIONS

5.1. Explain the importance of redundancy in mobile wireless core network infrastructure. How does distributing the core element functionality both logically and geographically improve system reliability and availability performance?

5.2. Compare the availability performance of the centralized versus distributed MSC. Assume that the MTBF of a single MSC component is 75,000 h and of the MTTR is 12 h and that the system serves a total of 500,000 subscribers across 10 MSCs (in the distributed case, assume uniformly distributed subscriber loading). Assume a redundant MSC configuration in the centralized case and single-thread MSCs in the distributed case.

5.3. A mobile wireless network is leasing backhaul capacity between its BTS locations and its centralized core. The backhaul network provider is proposing an SLA with a per circuit restoral maximum of 12 h and an annual aggregate availability of 99.8%. Assuming that the BTS equipment has an MTBF of 80,000 h, what MTTR should be specified to ensure that the network performance of the BTS does not diminish the backhaul service performance?

5.4. A certain region of a mobile wireless network has overlapping coverage for two BTS sites (assume 50% of the two regions overlap). Calculate the availability of the mobile station service for each BTS individually (use the MTBF and MTTR from Q5.3) and the improvement obtained in the overlapping region. If the location of a mobile station is uniformly distributed within the coverage area, calculate the total service availability within the region of coverage.

5.5. A mobile wireless network operator provides service in 50 cities throughout a geographic region. If the 50 cities of service are served by eight distributed core NSS units (with a designed core availability of 99.99%), calculate the availability impact of utilizing a centralized spares depot versus a distributed sparing plan. Assume that centralizing the spares location increases the MDT by 24 h and that the MDT for an NSS with on-site spares is 8 h.

6

TELECOMMUNICATIONS FACILITIES

Telecommunications facilities constitute a significant investment of capital. Building structures and systems require attention to the specific role of the facility within the communications system and the reliability and availability targets of that facility. Design details for a submarine cable landing station, a satellite teleport, a mountaintop microwave repeater site, or a central office switching center might all have very different architectural and engineering details but may all have the same availability target. The facility systems that most frequently have a direct affect on availability or reliability performance are:

- primary power systems
- battery backup systems
- heating, ventilation, and air conditioning (HVAC) systems

Power generation, distribution, and backup systems are a critical component of all telecommunications systems. Without reliable, available power, the electronic equipment on which communications relies does not function. It is important to have clearly

Telecommunications System Reliability Engineering, Theory, and Practice, Mark L. Ayers.
© 2012 by the Institute of Electrical and Electronics Engineers, Inc. Published 2012 by John Wiley & Sons, Inc.

defined assumptions regarding power systems when analyzing telecommunications systems. Input power for facilities generally comes from one of two sources, that is, commercial power, generator power, or both. Facilities frequently employ relatively inexpensive commercial power as a primary power source and install one or more generator systems for power restoral in the case of a primary power failure. In addition to employing reliable, available input power, telecommunications facilities almost always implement some variant of battery backup. Battery systems serve two basic purposes in a facility. First, battery systems inherently work as current surge suppressing devices. The potential for service impacting power events due to low voltage conditions, current spikes, or other transient events is minimized by placing battery strings inline with the power distribution system. In addition to power conditioning, battery strings serve the obvious task of supplying power in the case of a primary power system failure. The availability (and reliability) of facility power can be significantly improved by using properly designed battery systems. HVAC systems are often overlooked in availability analyses of telecommunications systems. Air conditioners, cooling fans, and heaters all play an important role in the management of environmental conditions that are acceptable for operation of electronic equipment. Electronics are specified for operation within a given range of temperature and humidity. Failure of HVAC systems result in changes to these environmental conditions. The rate of change of environmental conditions is often relatively slow and modeling equipment failures due to HVAC system failures can be difficult. When performing availability or reliability analyses, each of these systems should be considered to ensure that weak points or design flaws are exposed.

6.1 POWER SYSTEMS

Power systems for telecommunications facilities can be delivered by a commercial power utility, on-site generators, or a combination of both utility and generator power. Availability of the primary power system within a facility is dependent on both the presence and the performance of each of these elements. Figure 6.1 shows five different configurations of primary power deliveries. In scenario 1, the primary power system is simply a single feed from a commercial power utility with no on-site generator backup. Scenario 2 shows a single commercial feed with on-site backup power generation. Scenarios 3 and 4 introduce dual (redundant) commercial power feeds and backup generator systems. Both of these scenarios improve the availability performance of the primary power system while scenario 5 represents the highest (practically) achievable performance for a primary power system in which redundant utility and backup generator systems are implemented.

6.1.1 Commercial Power Delivery

Commercial alternating current (AC) power delivery to telecommunications facilities is extremely important in populated and developed areas. The availability of commercial utility power can vary dramatically depending on the region, provider, and

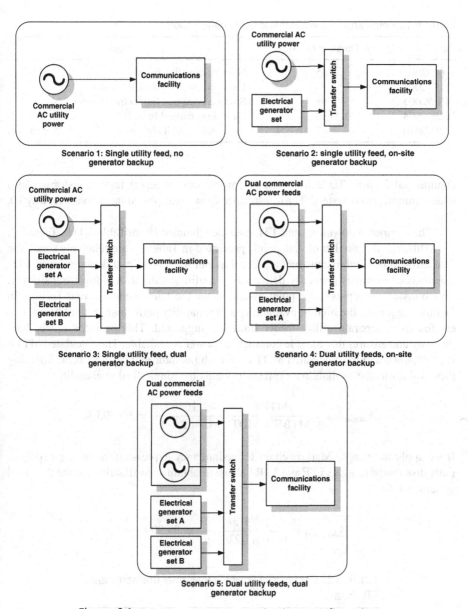

Figure 6.1. Primary power system redundancy configurations.

environmental conditions. When constructing a facility where commercial power will be used, it is a good practice to request an historical outage report from the utility. This data can provide valuable insight regarding frequency of outages, duration of outages, and so on. As an example, consider the case where a transformer feeding a tele-communications system has a track record for being susceptible to outages and may be considered to be the predominant weak link in the delivery of primary power to a

Table 6.1. Commercial Power Transformer Outage Report

Date	Duration (min)	Description
6/5/2002	145	High-voltage line failure
11/7/2004	35	Car accident caused switch failure
4/17/2005	98	Storm caused tree fall on transformer
10/28/2008	14	Snow load tripped breaker
12/1/2010	47	Tree contact with high-voltage line, breaker tripped

commercial facility. Table 6.1 shows a sample outage report for a transformer in a telecommunications system. The transformer shown was placed into service on July 10, 2001.

The computed downtime and TTF data are tabulated (from Table 6.1) in Table 6.2.

Although the number of data points presented in Table 6.2 are relatively small, the data can still be used to construct a general model for transformer downtime and TTF. Commercial power is generally delivered by a utility using a step-down transformer. This transformer represents a point-of-failure for the telecommunications facility. In facilities requiring the highest availability and reliability performance, dual transformers fed from separate utility feeder lines are suggested. This provides power feed diversity and ensures that reliable commercial power is available. The calculated MTTF is of 16,469 h and the calculated MDT is of 1.1 h (as indicated in Table 6.2). Applying these values to the availability expression yields the single feed availability:

$$A_{\text{single feed}} = \frac{\text{MTBF}}{\text{MTBF} + \text{MDT}} = \frac{16469}{16469 + 1.1} = 99.993\%$$

If we apply the simple Markov chain 1:1 redundancy expression (assuming exponentially distributed random TTF and TTR values) to obtain an availability for the dual feed system, we find

$$A_{\text{dual feed}} = \frac{\mu^2 + 3\lambda\mu}{(\lambda + \mu)(2\lambda + \mu)} = 99.999999\%$$

Table 6.2. Commercial Power Transformer Downtime and TTF Data

Event	TTF (h)	Downtime (h)
1	7,920	2.4
2	21,264	0.6
3	3,864	1.6
4	30,960	0.2
5	18,336	0.8
Mean	16,469	1.1

where $\mu = 1/\text{MDT}$ and $\lambda = 1/\text{MTTF}$. The extremely high availability performance obtained in the dual feed scenario assumes complete independence between the first and second utility feeds and independent diversely routed low-voltage drop connections both that cannot be affected by the same event. Transformers should be placed in a geographically diverse configuration such that a catastrophic event does not cause an outage of both transformers.

With sufficient outage data, a statistical distribution fit can be performed that estimates the outage duration and frequency. These statistical parameters can then be used as inputs in a Monte Carlo simulation that approximates the power availability of a particular commercial power utility. Monte Carlo simulation produces more detailed analyses that can place bounds on the availability performance (rather than simply providing average performance metrics). Figure 6.2 shows a Weibull distribution curve fit for the TTF and DT provided in Table 6.2.

Monte Carlo simulations are appropriate for power system modeling when more complex topologies that include battery backup and standby power generators. These configurations are discussed in the following sections.

6.1.2 Generator Systems

In many circumstances, system design dictates the necessity for standby or backup power generation using on-site diesel, natural gas, or another more exotic generator technology.

Each individual circumstance dictates a different power system design. In the case of insufficient utility availability performance, the mean downtime metric might be such that the duration of expected outages exceeds battery capacity and thus requires on-site power generation in order to achieve the desired availability performance. In cases where commercial utility power is not present (e.g., on a microwave mountaintop repeater site), the generator design must be selected to ensure that the power delivery is both highly reliable and highly available. Dual (redundant) generator systems are often employed in those cases.

The reliability and availability models of generator systems presented here assume that the generator has achieved steady-state operation (and thus the failure rate is constant). Generators are rotating machine devices and as such the constant failure rate assumption often adopted for solid-state electronic components is not always appropriate. In order for the constant failure rate assumption to hold, the generator system must be assumed to operate in a steady-state (bottom of the bathtub curve) condition. Rotating mechanical machines require break-in periods and frequent maintenance and experience wear-out. The study and analysis of generator reliability and availability is a well-documented field beyond the scope of this book. Readers interested in detailed or advanced treatments of this topic are encouraged to examine *493-2007: IEEE Recommended Practice for the Design of Reliable Industrial and Commercial Power Systems* (Hoopingarner and Zaloudek, 1998). One item of particular note with respect to telecommunications systems is the starting of generators. It has been documented in the Mission Critical West UPS Application Paper (DeCoster, 2010) that a failure to start is the most common cause of failure in a generator system. The TTR for a power failure

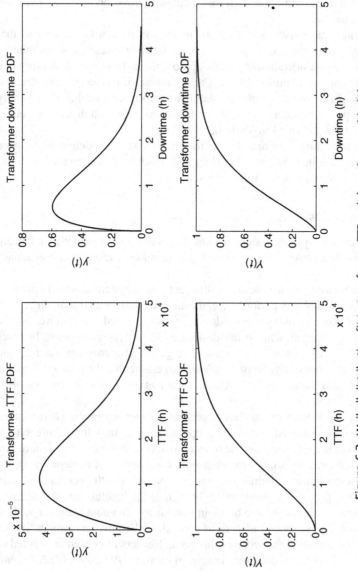

Figure 6.2. Weibull distribution fit to transformer TTF and downtime empirical data.

due to failure to start is almost always equal to or greater than the TTR for a system failure. Thus, if a backup generator fails to start on command, the opportunity for restoral of power and downtime avoidance is almost always lost.

Availability calculations involving rotating machine equipment follow the same procedure as discussed previously. The MTBF and MTTR of the generator are determined by modeling the TTF and TTR performance of the generator system while considering the time-dependent failure rate of the generator (if necessary). These system metrics are then used to assess the total generator system availability.

For the purposes of telecommunications systems, generators are most frequently deployed in one of three configurations: single thread, cold standby, and load sharing. Generator systems can be operated as a prime power supply or as a backup system that is called into service in the case of a commercial power failure. When operated as a prime power supply, the generator system is assumed to be running in a steady-state condition for availability and reliability performance analysis. When operated as part of a redundancy system, the generator design (by necessity) must include an automatic transfer switch (ATS) and an autonomous generator command start control system in order to achieve reasonable model performance. Failure to include these critical system components can result in analyses with inaccurate, optimistic performance expectations. Transfer switching (routing of system load from commercial power to backup generator power) and generator starting represent two common failure modes experienced in generator systems.

6.1.2.1 Single-Thread Generator Systems. The single-thread generator system consists of a single generator set and the associated electrical equipment required to route electrical power from the generator to the distribution system within the telecommunications facility. Figure 6.3 shows a simple sketch of the single-thread generator system.

Consider a generator set constructed for use in a village environment designed to provide power to a telecommunications facility. Such a village might be very remote and not connected to a national or provincial power grid. This generator set provides prime power (continuous) to the facility and is supplemented by 8 h of battery backup power. Failures of the generator must be resolved within the 8-h battery backup window in order to ensure that the power system continues to remain "available." Modeling of systems that utilize battery backup is only practically possible using Monte Carlo simulation methods. In this case, our model is constructed as shown below.

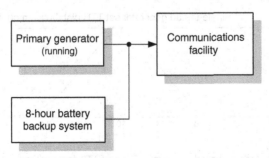

Figure 6.3. Single-thread generator system block diagram.

1. Model generator for system life-cycle duration using generator TTF distribution model.
2. Model repairs of generator system using TTR model.
3. In cases where repairs occur, examine the TTR for each instance and determine if TTR is more than battery capacity (BC).
4. In cases where the battery capacity is exceeded, the outage duration is equal to

$$\text{Outage} = \text{TTR} - \text{BC (h)}$$

The TTF for the generator set can be modeled using the methods described above for rotating machinery. The complexity of this model is dependent on the system analysis requirements. TTR model complexity will also be dependent on the model's expected output. In most cases, relatively simple TTF and TTR models will produce insightful analysis results. Figure 6.4 shows the assumed TTF and TTR models used in the simulation presented here. The generator set model assumes exponentially distributed TTF values with mean $\mu = 15,000$ h. Repair of the generator set is modeled by a normal distributed random variable with mean and standard deviation $\mu = 10$ h and $\sigma = 2$ h, respectively. Battery backup capacity is designed to ensure that some generator failure outages are bridged by the battery system. By observing the TTR CDF, we find that approximately 84% of the outages will have a duration longer than the battery backup capacity. This Monte Carlo simulation assumes a properly maintained generator set operated over a 10-year period.

This life-cycle period assumes that the generator does not experience any increase in failure rate early on (due to infant mortality) and has not begun to wear out by the end of the 10th year (this may not be a reasonable or accurate assumption). Note that in

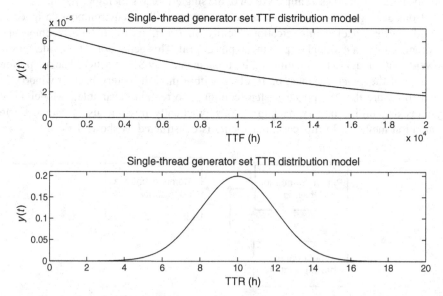

Figure 6.4. Single-thread generator TTF and TTR for a village environment.

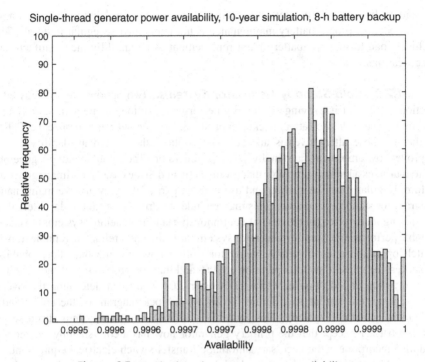

Single-thread generator power availability, 10-year simulation, 8-h battery backup

Figure 6.5. Single-thread generator system availability.

Figure 6.5, the y-axis is truncated to show samples of the simulation that did not result in 100% availability.

The mean availability achieved by the system is 99.985%. It should be no surprise that increasing the battery backup capacity will increase the availability performance of the generator system (approaching 100% as the probability of the TTR exceeding the battery capacity approaches zero). If we calculate the nominal availability (in the absence of battery backup) of the generator set, we can see that there is a decrease in availability performance (performance implications of battery backup systems are discussed in Section 6.1.4). The nominal availability of the generator set (without battery backup) is lower than the system utilizing a battery backup system. Selection of battery backup capacity and MTTR are clearly coupled to the availability performance and should be considered carefully, particularly in single-thread systems where availability performance is completely dependent on generator reliability, MTTR, and battery backup capacity.

$$A_{genset} = \frac{MTBF_{genset}}{MTBF_{genset} + MTTR} = 99.96\%$$

The availability calculated above reflects a single-thread generator set with no battery backup capacity. Decreasing the TTR will also improve the availability performance of the generator system by further ensuring that outages are more likely to be bridged by the battery backup system. Battery system failures are not modeled in this simulation but can occur if proper battery maintenance is not performed at regular intervals. This includes load testing and battery plant replacement as required by the manufacturer's specifications.

6.1.2.2 Cold-Standby Generator Systems.

Two options were discussed in Section 6.1.2.1 for improving availability performance of the on-site generator system. The first was to increase battery backup capacity and the second was to reduce the TTR. If neither of these alternatives is an acceptable option, the system availability can be improved by employing a standby backup generator. The "cold-standby" generator system utilizes a backup generator that is only activated in the case of a primary generator failure. Regular machine cycling and use is often part of the preventative maintenance program for standby generators. This may include start/run cycles at regular intervals or alternating load cycles (in the case of continuously running generator systems) to ensure reliable performance. Standby generator systems are inherently reliant on reliable transfer switch operation and on reliable generator starts. If we assume that the probability of successful transfer switch operation is 99% and that the probability of the generator starting when called upon is 99.5% (DeCoster, 2010), we can determine the overall generator system availability. Figure 6.6 shows a block diagram of the cold-standby redundant generator system. This system is identical to the system discussed in Section 6.1.2.1 except that the primary generator now has a cold-standby redundancy and also incorporates the necessary automatic transfer switch electrical equipment.

The Monte Carlo simulation methodology is modified from the single-thread case as shown below.

1. Model generator for system life-cycle duration using generator TTF distribution model.
2. Model repairs of generator system using TTR model.

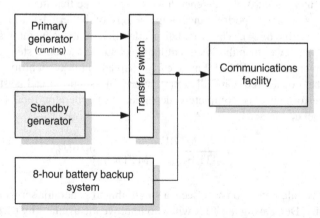

Figure 6.6. Cold-standby redundant generator system block diagram.

3. In cases where the primary generator fails, determine if transfer switch operation is successful and if the backup generator starts.

4. In cases where the backup generator fails (transfer switch failure or failure to start), examine the TTR for each instance and determine if TTR is more than BC.

5. In cases where the battery capacity is exceeded and the backup generator fails, the outage duration is equal to

$$\text{Outage} = \text{TTR} - \text{BC} \ (\text{h})$$

Failure of the primary generator in this system only results in a power outage if the backup generator fails (due to a transfer switch failure or a failure to start) and the battery life is exceeded. Assuming the same 10-year system life assumptions presented in the single-thread case, we can calculate the new generator system availability. In this case, the presence of a backup generator improves the average system availability to 99.9998%. This is a significant increase in availability performance over the single-thread system design. One item of note is the significant increase in life-cycle simulations for which no failures occur. In the single-thread case, the results showed that less than 1% of simulations experienced failure-free operation. In the cold-standby redundant case, this number increases to approximately 93%. Repairs will be necessary in many life cycles, but there is now a real, measurable probability of failure-free operation.

If the desire of adding a redundant generator is to reduce the battery capacity or increase the MTTR, both of these options can be optimized while designing the system to meet a target availability metric. Figure 6.7 shows a histogram of the achieved life-cycle

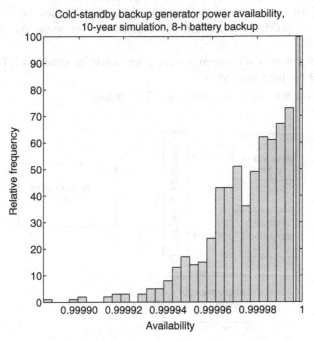

Figure 6.7. Cold-standby redundant generator system availability.

availability for each simulation. The average availability and percentile statistics discussed above are calculated from the population performance statistics produced in the Monte Carlo simulation. The y-axis of the figure was truncated to demonstrate the distribution of availabilities for life cycles not experiencing 100% availability.

6.1.2.3 Load-Sharing Generator Systems.

Failure to start represents the single most significant failure mode in redundant generator system design. The critical time period for power restoral is quite small (equal to the battery capacity) and is almost always less than the TTR. This means that a failure to start will almost always result in a power outage (and ultimately unavailability) on the generator system. One method for mitigating the failure to start scenario is to deploy generators configured for "load-sharing." In this configuration, two (or more) generators are continuously operated and the output of each generator is connected to paralleling equipment. This paralleling equipment effectively bonds or sums the outputs of the individual generators into a single feed for use by the telecommunications facility. Load-sharing generator systems are frequently more expensive and technically more complex than standby redundant systems. The advantage of a load-sharing system is that because all generators are always running, the system is not reliant on a latent generator starting when called upon for service. The load-sharing system is designed such that the failure of any one (or more) generator does not result in a power outage or brownout condition. Figure 6.8 shows a block diagram of a dual generator load-sharing system. This system utilizes the same generator sets presented in the single-thread and cold-standby models but also employs the required paralleling equipment for summing the output of the two generator sets.

Monte Carlo simulation of the load-sharing system follows the same procedure used in both the previous models but with slight variations to the system conditions.

1. Model generator for system life-cycle duration using generator TTF distribution model for both generators.
2. Model repairs of a generator using TTR model.

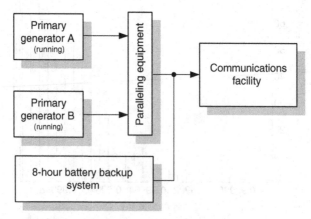

Figure 6.8. Load-sharing generator system block diagram.

3. In cases where two generator failures occur, examine the TTR for each instance and determine if TTR is more than BC.

4. In cases where the battery capacity is exceeded, the outage duration is equal to

$$\text{Outage} = \text{TTR} - \text{BC (h)}$$

This system provides the highest availability performance since, once in operation, it is not reliant on load transfer switching or machine starting to provide redundant operation. Figure 6.9 shows the results of a Monte Carlo availability simulation for a 10-year life cycle using two identical generator sets with the TTF and TTR models shown in Section 6.1.2.1. The generator system life-cycle simulation is performed 2500 times in order to produce the histogram in Figure 6.9. In this figure, the y-axis has again been truncated to demonstrate the distribution of availabilities for simulations that did not result in 100% availability. The average availability achieved in this simulation is 99.999998%. Clearly, the target MTTR, battery capacity requirement, or both can be relaxed to achieve a more reasonable target availability. If we target the 99.9998% availability achieved in the cold-standby generator case, we find that we can relax the MTTR to 25 h while still achieving the desired target performance with 8 h of battery backup. Using these values, we achieve an average availability of 99.9998% for the system. Figure 6.10 shows the repair distribution used in the revised MTTR model.

As a result of the modified simulation parameters, the battery backup provides little value with regard to power availability performance. The value of the battery backup

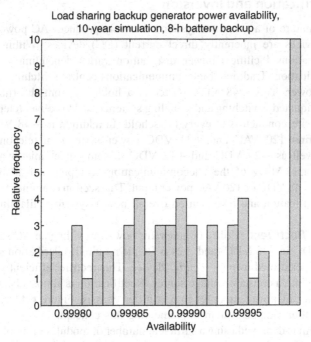

Figure 6.9. Load-sharing generator system relaxed TTR model.

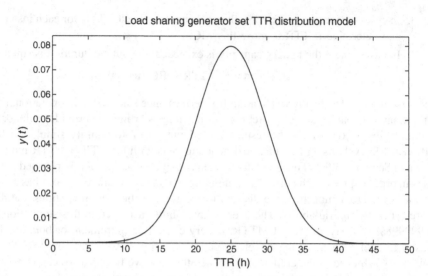

Figure 6.10. Load-sharing generator system availability.

system in this configuration comes from the capacitive filtering effect and the ability to survive input power surges and brownout conditions.

6.1.3   Rectification and Inversion

The rotating nature of a generator system inherently produces AC power. Solid-state electronic devices are inherently direct current (DC) devices. Within the field of telecommunications facilities, power distribution varies depending on the facility type and application. Traditional telecommunications central switching offices utilize −48 VDC power. This −48 VDC power is a holdover from the (now becoming somewhat antiquated) switching and signaling systems used to deliver telephone service over copper wire conductors to every household. In addition to −48 VDC, it is also common to find 120 VAC and +24 VDC power systems in telecommunications facilities as well as −24 VDC and +12 VDC in some older microwave and VHF repeater systems. Many of the telecommunications equipment available today are having either −48 VDC or 120 VAC power input. This section examines the availability impact of rectification and inversion on overall power system availability.

6.1.3.1   Rectifiers. Utility or generator power providing an AC source must be converted to DC in −48 VDC (and +24 VDC, if required) distribution systems. This conversion is performed using a *rectifier* device. This rectifier efficiently converts the AC waveform into a constant voltage source. Rectification is almost always performed using a number of power modules in a rectifier chassis. Figure 6.11 shows a block diagram of a four-module system implementing this concept.

The system is deployed using a sufficient number of modules to supply the required current. Most systems are designed to support additional modules as the current

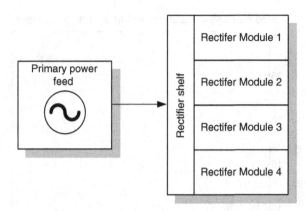

Figure 6.11. Modular rectifier system block diagram.

capacity requirement grows. Availability analysis of the system shown can be done for one to four modules. Two types of designs can be envisioned to implement the modular design. The first concept is a 1:N redundancy system where a single standby backup module protects up to N active modules. The protection module in this scenario is a "cold-standby" unit and does not supply current until commanded into operation. The second system design is analogous to the "soft-fail" system described in Chapter 4 (or as described in Section 6.1.2.3). In this system, all the N modules are supplying current (load sharing) at all times. The capacity of the N modules to provide current is designed such that the failure of any one (or more) module can be absorbed by excess capacity in the other modules. Figure 6.12 shows a graphical depiction of the two scenarios.

For the purposes of analysis, we will assume a fully populated four-module rectifier shelf. A 1:N system consists of three active rectifier modules and a single out-of-service protection rectifier module. If we assume that the shelf availability is dependent only on the failure and repair rates of the rectifier modules, we can calculate the availability of the rectifier using a three out of four availability model. The difference in the 1:4 and soft-fail availability models is subtle. In the 1:N model, the protection module is cold standby and is not accumulating hours toward its failure while offline. In the soft-fail model, all four modules are online all the time. The results shown below (in Figure 6.13) provide

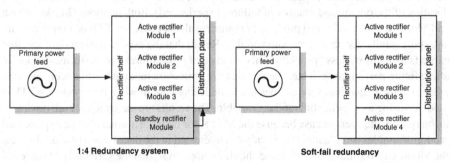

Figure 6.12. 1:N and soft-fail rectifier design descriptions.

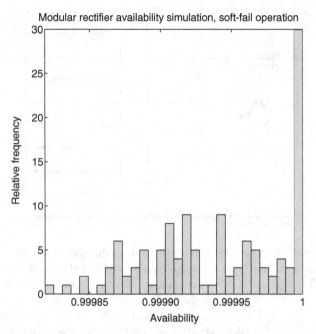

Figure 6.13. Soft-fail rectifier system availability distribution.

availability distribution for the soft-fail system. Each rectifier module in the simulation is assumed to have an MTBF of 50,000 h with an exponentially distributed TTF. The repair of the 1:N system is modeled using a Markov chain process. The repair of the soft-fail rectifier system is modeled using a normal distributed random process with a mean of 24 h and a standard deviation of 5 h. Even with relatively low MTBF performance (5000 h), the redundant system still achieves very good availability performance. In both the 1:N and soft-fail cases, the average life-cycle availability achieved is 99.9998%. The Monte Carlo simulation results show that approximately 98% of the 20-year life cycles are failure free. Only in 2% of the cases does a dual rectifier module failure occur (this comment refers to the distribution of simulations experiencing performance with an availability of less than 100%, the performance experienced in that particular life cycle is dependent on the duration of the outage and number of failures experienced). Both methods (Markov chain and Monte Carlo simulation) produce the same results because the TTF is modeled using an exponentially distributed random variable. Recall that the exponential random variable exhibits a "memory-less" property. This means that past failures and performance do not affect future performance. Thus, the fact that the protection module has not been operated prior to being called into service does not provide any benefit when this module's TTF is modeled using an exponential random variable. The results are the same for both modeling techniques in the average case because the MTTR for both the normal and the exponential random variables is the same. If the analyst is interested in the distribution of availabilities, the Monte Carlo simulation technique should be applied so that the variability of achieved availability (dependent on the TTR random variable distribution) can be assessed. Note that the histogram shown in Figure 6.13 is intended to show the variability of availability for instances not achieving an availability of 100% for a particular lifecycle. Mean and standard deviation of the availability performance is computed using the population values obtained in the Monte Carlo simulation.

6.1.3.2 Inverters. Facilities utilizing DC power distribution sometimes operate equipment that is available with AC-only power input capabilities. In these cases, two options exist. If battery backup is not required, for example, with noncritical equipment, the equipment can be powered directly by a branch circuit feed from the facility main AC power feed. In cases where battery backup is required, two options exist. The first option is to install and operate an uninterruptable power source (UPS). This option is discussed in Section 6.1.4. A second alternative is to install an AC inverter device. The AC inverter converts the DC power source to an AC supply suitable for use with the equipment. AC inverter systems are frequently implemented in telecommunications systems using the same modular approach described in Section 6.1.3.1.

6.1.4 Battery Backup Systems

Battery systems are an integral part of all telecommunications facilities. Battery systems are used to provide surge suppression and uninterrupted power in the case of a primary power failure. Historically, the battery system within a central telephone office facility provided line voltage (-48 VDC) for providing loop current on a copper telephone line. Today, these loop-powered analog telephone lines are still very much in use (although in some cases, they are being replaced by new digital telephony systems). Central telephone offices still house large -48 VDC battery plants to power central office switching systems. These line cards provide the DC loop current power for analog telephone lines throughout a city. Standard practice (with its roots in the Bell System and regulatory requirements) demands a large central office backup battery plant for providing reliable, lifeline telephone service in the event of primary AC power system failure. Because of the long and continuing history of telephone operators using -48 VDC in their central offices, most telecommunications equipment has been designed to operate on this supply voltage and to be connected to central DC distribution systems. Although the use of switching center sourced line power for telephone set ringers is all but extinct, the requirement for the presence of a battery backup system remains. Facilities utilize battery backup systems to provide the uninterrupted power required in both AC and DC supply environments. Although the power provided by a battery backup system is always DC in nature, the system used to provide the battery capacity is referred to by one of two names. In DC distribution systems, the battery backup is provided by what is commonly referred to as a "battery plant." AC systems utilize an Uninterruptable Power Supply (UPS). Both systems are implemented as a number of cells that are interconnected in serial and parallel circuits in order to build a system capable of the required current and voltage. Figure 6.14 shows a block diagram of a -48 VDC system made up of four strings of 2 VDC cells (where each string consists of 24 individual cells) that are combined to provide a battery plant with a rated capacity (in ampere-hours) at a nominal voltage of -48 VDC.

48 VDC
(24 cells @ 2 VDC each)

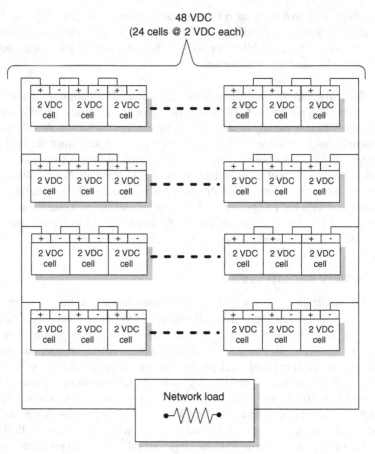

Figure 6.14. −48 VDC battery plant block diagram.

6.1.4.1 Battery Technologies. Telecommunications facilities implement battery backup using one of a number of battery technologies. The battery technology used is dependent on the facility size, battery cost, and required battery performance. The discussion below is intended to provide the reader with a very basic knowledge of the most common battery technologies used in telecommunications systems. Readers interested in obtaining a more comprehensive understanding of the pros and cons of each battery technology are encouraged to research the extensive academic and industry knowledge bases on this topic.

A common battery technology used in telecommunications facilities is the valve-regulated lead–acid (VRLA) battery. These batteries are sometimes referred to as "gel-cell" or "absorbed glass mat (AGM)" batteries because the electrolyte solution is either dissolved in a silica powder (forming a gel) or in a fiberglass mat. The purpose of both of these substrates is to immobilize the electrolyte. VRLA batteries have a high power-to-space ratio, do not require electrolyte addition, and are not "spillable." In addition to being physically stable, the oxygen and hydrogen produced by the battery discharge process recombines to produce water. Only in cases where the gas production rate exceeds the recombination rate will electrolyte leakage occur in VRLA batteries. Regular electrical performance testing (internal impedance) must be performed to ensure that battery performance has not degraded due to electrolyte leakage. Faulty or failing cells must be replaced promptly to ensure that the battery plant capacity is maintained at a specified level. VRLA batteries can fail prematurely due to a number of different causes. These causes include elevated operating temperature, frequent discharge or recharge cycles, and poor installation practices. In order to achieve the maximum battery life performance, the VRLA battery cells must be maintained at temperatures equal to or less than 77°F. Discharge or recharge cycles should be designed such that the depth of discharge (DoD) is minimized for the battery plant. Designing the battery plant capacity so that the DoD does not result in a battery system degradation is a good practice to maximize battery performance and maintain system life. VRLA batteries are subject to a slow loss of capacity over time. The internal resistance of the battery increases as the internal conductive paths age and as a result of this increased resistance, the current sourcing capacity of each battery cell decreases. Failures due to capacity reduction are often realized by a reduction in a battery plant's ability to supply a current for a specified period of time. Cell aging effects are detectable by regular measurement of battery cell internal resistance. A poorly maintained battery plant can cause a catastrophic failure in the case that the VRLA battery electrolyte is allowed to completely evaporate. In this circumstance, the cell is effectively an open circuit and no current is allowed to flow. Importantly, all of the causes of premature failure indicated above are avoidable using proper operational and maintenance procedures. Early detection and replacement of faulty VRLA cells is critical to reliable operation of a VRLA battery system. Reliability models for VRLA systems have been developed (Koizumi and Yotsumoto, 2011) that approximate the early failure and manufacturing defects present in current VRLA technology.

Flooded (or vented) lead–acid batteries are also used in large-scale telecommunications facilities. The flooded lead–acid battery is physically very similar to the VRLA AGM battery except that the lead plates within the battery are surrounded by (flooded) a liquid electrolyte solution. Flooded lead–acid batteries tend to be very heavy and are somewhat inefficient but can provide very large surge currents on demand. In addition, the venting of electrolyte solution produces oxygen and hydrogen gas that must be removed from the facility using appropriate air-handling measures. Properly maintained flooded lead–acid batteries have very long service lifetimes and very good performance.

In addition to the traditional VRLA and flooded lead–acid battery technologies, a number of emerging battery technologies are appearing in telecommunications facility design. Lithium-ion battery technology offers lightweight, high power and energy density, and low space requirements. Safety, longevity, and price are all relevant concerns with lithium-ion technology. As the lithium-ion battery technology matures, telecommunications facilities will likely see its widespread use.

6.1.4.2 Battery Capacity. The impact of battery capacity on overall system availability was discussed briefly in the context of generators and primary power systems. The downtime or TTR of a network facility must be considered when designing the battery backup capacity. As the battery backup capacity increases with respect to the TTR, the probability that the facility power system is unavailable becomes zero. We can calculate the probability of the battery capacity being exceeded for a given MTTR and the impact on a system's availability. Consider a system (to be battery backed-up) with a normal distributed TTR with a mean equal to 10 h and a standard deviation of 2 h. The PDF and CDF of this TTR distribution are shown in Figure 6.15.

A range of battery capacities and the associated probability of that capacity being exceeded (for a normal distributed TTR) are calculated below in Table 6.3. As shown in this table, the given normal distributed random variable selecting a battery capacity equal to 150% of the MTTR will result in almost 100% of outages being bridged by the battery system. The impact of this probability will vary depending on the type and complexity of redundancy implemented on the system. The results of each battery capacity on the single-thread and cold-standby generator systems are shown. In the single-thread case, the availability achieved by selecting a battery capacity equal to or greater than 100% of the MTTR results in a significant improvement in availability performance. The cold-standby redundant case experiences a similar performance improvement to the single-thread case. In both circumstances, implementing a battery system with capacity equal to or exceeding the MTTR will result in an order of magnitude increase in availability performance.

Figure 6.16 shows the values given in Table 6.3 in a graphical form. Note that both curves share the same general linear improvement until the capacity reaches 100% of the MTTR value after which the improvement diminishes for increased battery capacity.

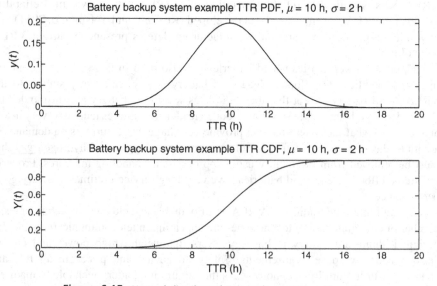

Figure 6.15. Normal distributed TTR with $\mu = 12$ h and $\sigma = 3$ h.

Table 6.3. Battery Capacity Versus Availability Performance for Single-Thread and
Cold-Standby Generator Systems

Battery Capacity (% of MTTR)	Probability of Battery Capacity Exceeded (%)	Single-Thread Generator Availability (%)	Cold-Standby Generator Availability (%)
0	100	99.94	99.9990
25	99.99	99.95	99.9993
50	99.4	99.97	99.9995
75	89.4	99.98	99.9998
100	50.0	99.994	99.99991
125	10.6	99.999	99.99997
150	0.6	~100	~100

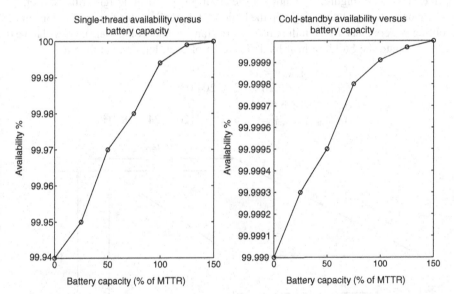

Figure 6.16. Availability performance versus battery capacity for single-thread and cold-standby generator systems.

6.2 HEATING, VENTILATION, AND AIR CONDITIONING SYSTEMS

Control of facility environmental conditions is crucial to the proper operation of tele-communications equipment. All telecommunications equipment produces heat as a by-product. This heat is evacuated from the equipment by convective or forced air cooling. Air heated by the equipment is cooled in the facility by air exchange methods. Computer room air conditioners (CRACs) are frequently used to provide the room cooling capacity required to sustain a specified ambient temperature within the environment in the presence of the equipment heat sources. A number of methods are used to deliver cool air to the equipment intake including raised floors, cold isle containment, and free air cooling. In poorly insulated equipment shelters, heating methods may be required in cold climates. All telecommunications equipment is designed for operation within specified environmental conditions. These conditions include temperature ranges (high and low limits) and humidity ranges (high and low limits). Failure to ensure that the environment is controlled within the specified ranges will result in eventual equipment failure.

HVAC system failures have an indirect relationship with the service availability of a telecommunications network. Even in the case of a complete environmental control system failure, the temperature and humidity conditions will not change instantaneously. Rather, the heat load and humidity differential (between the cooling or heating capacity and the equipment heat output) will cause the room temperature to rise (or fall) at a computable rate. Consider a communications shelter measuring 12 feet wide, 18 feet long, and 10 feet high. The shelter is constructed using wood 2×6 framing and is covered using a fiberglass shell. Figure 6.17 shows a sketch of this physical design of the shelter.

In order to simulate changes in the building temperature due to perturbations in the cooling system (such as the failure of a CRAC, fan, or other component), we will have to first calculate the building heat load. The area of each building surface is

$$A_{\text{left and right wall}} = 18 \times 10 = 180\,\text{ft}^2$$
$$A_{\text{floor and roof}} = 18 \times 12 = 204\,\text{ft}^2$$
$$A_{\text{door}} = 3 \times 8 = 24\,\text{ft}^2$$
$$A_{\text{front and back wall}} = 12 \times 10 = 120 - 24 = 96\,\text{ft}^2$$

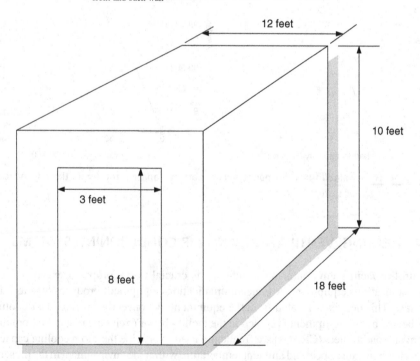

Figure 6.17. Fiberglass communications shelter dimensions.

Assuming that the structure is insulated equally in all dimensions, the heat gain (or loss) of the building can be calculated (assuming a set-point temperature inside the building of 18°C and outside temperature of 24°C and R-19 wall insulation):

$$Q = A \times U \times (t_{interior} - t_{exterior})$$

$$Q = (180 + 204 + 96) \times \left(\frac{1}{19}\right) \times (10.8) = -273 \, \text{BTU/h} = -80 \, \text{W}$$

where the temperature differential of 6°C has been converted to Fahrenheit (10.8°F). This is the overall heat gain of the building with no interior equipment in place. If we assume an interior heat load of 2.5 kW, we can calculate the required cooling capacity of the building to support the stated heat load. The cooling capacity of CRAC units is specified in terms of *tons* (1 ton is equal to 3530 W). The air conditioning requirement of the system described above is

$$\text{CRAC size} = \frac{2.5 \times 10^3 \, \text{W}}{3530 \frac{\text{W}}{\text{ton}}} = 0.7 \, \text{tons} \rightarrow 1 \, \text{ton}$$

An air conditioning system capable of supplying 1 ton of cooling capacity is required to ensure that the heat produced by the telecommunications equipment can be evacuated in a steady-state condition.

We will examine two different cooling solutions, one in which a single air conditioner unit is used and a second where redundant air conditioners are used to cool the room.

In the first case, a single air conditioner is used. Upon the failure of the air conditioner, net heat load inside the room goes up from nominally zero (since the air conditioner is evacuating all of the heat to maintain the interior temperature set point) to a value 2.5 kW higher. The temperature of the room will continue to rise in the absence of any air flow to remove the heat and the equipment will overheat. Assuming an air volume within the space equal to the $L \times W \times H$ times an airspace factor of 60%:

$$V_{shelter} = 12' \times 10' \times 18' \times 0.6 = 1296 \, \text{ft}^3$$

The airspace factor represents that 60% of the space inside the shelter is occupied by the equipment. Assuming that the heat capacity of air is 0.02 BTU/ft^3 · °F, we calculate the heat capacity of the shelter airspace to be

$$H_{shelter} = 0.02 \times 1296 = 25.9 \, \text{BTU}/°\text{F}$$

If we convert the heat load from Watts to BTU/s, we find

$$Q = 2.5 \times 10^3 \, \text{W} \times \frac{3.41 \frac{\text{BTU}}{\text{h}}}{1 \, \text{W}} = 8.5 \times 10^3 \frac{\text{BTU}}{\text{h}}$$

We can calculate the heat loss of the building due to the increasing temperature differential between the interior space and the outside air (based on wall insulation with an R-value of 19):

$$H_{walls/ceiling} = (180 + 204 + 96) \times \left(\frac{1}{19}\right) \times \Delta t = 25.3 \Delta t \left(\frac{\text{BTU}}{\text{h}}\right)$$

We can calculate the rate of temperature increase based on the 2.5 kW heat load as

$$r_1 \left(\frac{°F}{s}\right) = \left(8.5 \times 10^3 \frac{BTU}{h} - 25.3\Delta t \frac{BTU}{h}\right) \times \frac{°F}{25.9\,BTU} \times \frac{h}{3600\,s}$$

In the case that the system is utilizing two (soft-fail) air conditioners, the temperature of the room will continue to increase but at a lower rate. Assuming that the heat load is halved by retaining a single functioning air conditioner

$$r_2 \left(\frac{°F}{s}\right) = \left(4.25 \times 10^3 \frac{BTU}{h} - 25.3\Delta t \frac{BTU}{h}\right) \times \frac{°F}{25.9\,BTU} \times \frac{h}{3600\,s}$$

Figure 6.18 plots a graphical depiction of the rates shown. The horizontal line in Figure 6.18 shows that at 135°F, the system will begin taking errors (and thus failing).

A more sophisticated model would show that failure occurs earlier in the system. The absence of cool air at the input of the equipment will increase the temperature more rapidly than shown here. A full treatment of building HVAC design is beyond the scope of this book. The purpose of the model shown is to demonstrate both the criticality of the design and the operational phases of HVAC systems in communications networks. HVAC systems frequently employ 1:N redundancy. In these systems, the failure of a single AC unit does not reduce the total system cooling capacity below the required performance level. The trade-off in systems using 1:N redundancy is cost. For systems

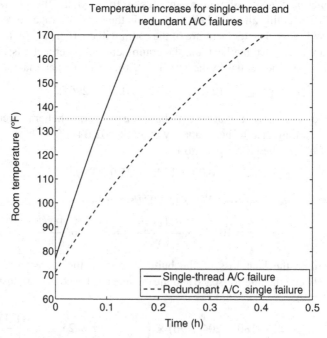

Figure 6.18. Room air temperature increase rate for two A/C scenarios.

using 1:1 or 1:2 redundancy, the costs of implementing redundancy can be very high. Clearly (as shown in Figure 6.18), the absence of redundancy can lead to rapid failure in the system.

QUESTIONS

6.1. Explain the challenges associated with modeling HVAC system availability performance and its impact on service availability. How can an HVAC system failure lead to a service failure?

6.2. Compute the TTF, downtime, MTBF, and MDT metrics for the empirical failure data tabulated below.

Date	Duration (min)
4/3/2006	79
1/19/2007	34
9/14/2007	8
3/22/2008	142
11/6/2008	58
5/31/2009	17
10/8/2010	92

6.3. Using the data in Q6.2, develop a Weibull distributed random variable model for the TTF and TTR. Plot the PDF of each random variable and provide their scale and shape parameters.

6.4. What is the most frequent cause of generator failures in a telecommunications facility? What design methods can be used to avoid this type of system fault?

6.5. A telecommunications facility is designed to provide -48 VDC power in a continuous manner to the connected equipment. The facility consists primary commercial power with battery backup and a standby generator system. If the battery system is designed to provide only enough capacity ensuring that the generator switching event is seamless, calculate the availability of the power system. Assume that the individual components have MTBF and MTTR values as shown below and that the transfer switch has a probability of successful operation equal to 99.5%.

Component	MTBF	MTTR
Commercial power	20,000 h	3 h
Battery backup	N/A	N/A
Generator set	15,000 h	6 h

6.6. If the system described in Q6.5 is reconfigured to provide redundant commercial power feeds (assume statistical independence in the failure of these feeds), what is the resultant availability?

6.7. In what configuration will redundant generator systems operating continuously (prime power) achieve optimal availability performance? Why is this the case? Implementing this configuration allows the designer to relax what system parameter(s)?

6.8. Battery capacity in a telecommunications facility is intrinsically tied to what operational metric? What is the approximate optimal battery capacity to ensure that the cost or benefit of deploying a battery system is achieved?

6.9. A telecommunications facility is designed to provide HVAC cooling using three out of four CRAC (1 ton each) system. Assuming that the MTBF of each CRAC unit is 25,000 h and the MTTR is 14 h, calculate the availability of the CRAC system. If the cooling requirement of the facility is 4 tons, what is the probability that a failure occurs for which the HVAC system is not sufficient to cool the environment? How long does it take for the system to reach a temperature of 135°F (starting at 65°F) assuming a $12' \times 12' \times 20'$ shelter (for simplicity, assume no heat transfer occurs between the air inside and outside of the shelter) in which 60% of the shelter volume is occupied by air and a heat load of 12.5 kW?

7

SOFTWARE AND FIRMWARE

The topic of software and firmware reliability modeling within the field of tele-communications systems is a challenging one. The underlying principles affecting software reliability are fundamentally different than those for hardware reliability. The metrics, behavior, and performance expectations for software and firmware all require a different approach than those used for hardware reliability.

In today's implementation of modern digital communications systems, telecommunications engineers use a range of computing architectures including digital signal processors (DSPs), field-programmable gate arrays (FPGAs), application-specific integrated circuits (ASICs), very large-scale integrated (VLSI) circuits, and/or COTS general computing processors. Production telecommunications systems use platforms ranging from proprietary, custom-developed hardware and firmware to off-the-shelf single-board computer systems adapted to a specific application. Reliable software, firmware, and middleware are fundamental components of a reliable telecommunications system. Although software reliability plays a crucial role in communications systems, its impact is not frequently modeled in reliability or availability analyses for production networks.

Frequent software revisions and firmware upgrades are commonplace in the field of telecommunications. Whether the purpose of the change is to add additional features or

Telecommunications System Reliability Engineering, Theory, and Practice, Mark L. Ayers.
© 2012 by the Institute of Electrical and Electronics Engineers, Inc. Published 2012 by John Wiley & Sons, Inc.

to resolve a software programming error (a "bug"), modification of software or firmware should be met with caution. The introduction of a new software or firmware revision can introduce new or previously corrected bugs into an otherwise stable system.

Most methods for modeling or predicting software reliability (and availability) are focused on development and testing cycles within the software vendor's facility. The customer rarely has access to any measure of expected or predicted software reliability performance through standard product research and documentation channels (such as the hardware MTBF or failure rate). Instead, customers experience a vendor's performance with regard for software bug fixes and feature development firsthand. This experience frequently comes only after suffering significant outages and cost.

7.1 SOFTWARE FAILURE MECHANISMS

Failures in software occur for reasons that are fundamentally different than those observed in hardware components. It is useful to compare the concepts for hardware reliability to a software system to build a better understanding of the differences.

7.1.1 Failure Causes

Failures within system hardware components are due to physical influences such as heat, humidity, solar radiation, manufacturing defects, and other conditions. Hardware components left to operate indefinitely will inevitably fail (typically due to mechanical, electrical, or thermal stresses), whereas a software component failure is due solely to design issues. Software reliability is not dependent on environmental conditions, manufacturing process, or time, in general, but is only dependent on its current operating state. Software entering an unknown or undefined state can cause a system failure. The commonly observed hardware reliability bathtub curve showing early failure, steady state, and wear-out periods of operation do not apply to software or firmware. Rather, in the software reliability case, the system is a function of changing failure rates as new revisions are adopted over time, bugs are corrected, and new features are added. As time increases, the system eventually enters obsolescence and a steady-state failure rate settles out for the system. This does not imply that the software is failure free but rather that its failure rate is no longer changing significantly with time. Figure 7.1 shows a sketch comparison of the classical bathtub curve and an example of the software failure rate curve described above, where each stair-step function is the result of loading a new software version that corrects past bugs and, in this example, shows that new releases most of the time introduce fewer new bugs than they correct.

7.1.2 Software Repair

Repair of hardware failures is a straightforward problem. The failed component in a system is identified, removed, and replaced. The most common assumption (associated with exponentially distributed random variable TTF models) is that the replacement of the failed component returns the system to an "as good as new" operational state. Failures

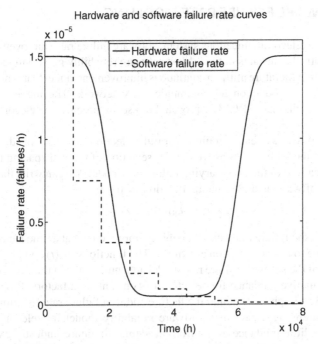

Figure 7.1. Sample hardware and software failure rate versus time curve comparison.

within a software component are more difficult to identify and can be equally difficult to resolve. Frequently, problems manifested by a software failure are indirectly related to the problem observed. Experienced technicians gather evidence and diagnose the problem within a telecommunications system. The solution to many problems is often a software restart. This restart operation places the software back into a known operational state from which its primary logic can start fresh. In some instances, restarting the software will temporarily correct the issue unless the software is being forced to an ambiguous or undefined state. Technicians collect the data gathered from the failure event and open a trouble ticket with the software vendor. Responsive vendors can produce software patches (likely without extensive regression testing) in a few days to correct issues if the cause of the software failure can be quickly identified (e.g., capturing the failure event with log files). In some instances, however, it can take months for a comprehensive software repair to take place and loading the new software to correct the "known" issue carries its own risks. Without proper regression testing and verification, the repaired software can introduce new, equally significant software defects.

Software, once deployed on a system, does not change without external intervention. External interventions can be due to either feature upgrades or bug fixes. In either case, the change to the software (or firmware) introduces a change in the overall software failure rate. For the purposes of a telecommunications system, the software changes can be classified into two categories: reliability improvements and feature additions or upgrades.

7.2 SOFTWARE FAILURE RATE MODELING

In the case of a telecommunications system, software failure rate improvements come at specified, controlled instants in time. Most software reliability growth models (SRGMs) present in current literature utilize a continuous improvement model that is not necessarily appropriate for a production telecommunications network. The model shown below modifies the traditional SRGM approach for use in production telecommunications systems.

Modeling real systems requires careful selection of each of the parameters governing the model behavior as well as the selection of interval spacing (for software change application) of the time-varying failure rate model. We can write the total failure rate for the software as a continuous function of time:

$$\lambda(t) = \lambda_{RI}(t) + \lambda_{FA}(t) \tag{7.1}$$

where $\lambda_{RI}(t)$ is the failure rate due to reliability improvement at a time t and $\lambda_{FA}(t)$ is the failure rate due to feature addition at a time t. The functions $\lambda_{RI}(t)$ and $\lambda_{FA}(t)$ governing the behavior of the software failure rate model are enumerated in this section along with some representative guidance for selection of constants and factors. Proper definition of the reliability improvement and the feature addition failure rate functions represents the most challenging aspect of a software reliability model for telecommunications systems. Very little guidance exists from academic literature, industry experience, or otherwise to assist engineers in developing software reliability models for practical, real telecommunications systems.

7.2.1 Reliability Improvement Failure Rate

Bug fixes or reliability improvements result in a software failure rate improvement (decrease) and may occur as part of regularly scheduled software updates or as point releases applied in an ad hoc manner. A mathematical expression for the reliability improvement failure rate function is given below. The expression shown takes on values for discrete ranges of time (from t_n to t_{n+1}) based on the initial failure rate value $\lambda_{RI\ Initial}$ and the improvement factor D_n.

$$\lambda_n = D_n \times \lambda_{n-1} \text{ for } n \geq 1 \text{ and } t_{n-1} \leq t < t_n \tag{7.2}$$

where $0 < D_n < 1$ and $\lambda_0 = \lambda_{RI\ Initial}$. The total expression for failure rate improvement is given by

$$RI = \lambda_{RI\ Initial} - \lambda_{Steady\ state} \tag{7.3}$$

where $\lambda_{Steady\ state}$ is the failure rate after reliability improvement software revisions no longer produce a significant improvement in software reliability. Determining the initial value of the software failure rate ($\lambda_{RI\ Initial}$) is problematic. It is unlikely that an analyst will have access to substantiating data that unequivocally support the initial value selected. Rather, it is more likely that the model will be based on a combination of professional experience and mathematical logic. In the case of a single device model

Table 7.1. Perceived Code Complexity Factor

Perceived Code Complexity	Complexity Factor (%)
Very low	25
Low	50
Medium	100
High	150
Very high	175

(such as a modem, multiplexer, or control panel), the analyst might apply the following logic to select a reasonable initial software failure rate value.

1. Compute the total device hardware failure rate using the methods presented previously in this book.
2. Assess the complexity of the software/firmware code base by examining the system logic at a high level. Assign the code base a complexity factor as shown in Table 7.1.
3. Compute the initial software reliability value by multiplying the steady-state hardware failure rate by the complexity factor assigned in step 2.

Thus, the initial failure rate value $\lambda_{\text{RI Initial}}$ can be written as

$$\lambda_{\text{RI Initial}} = \lambda_{\text{hardware}} \times \text{CF} \qquad (7.4)$$

where $\lambda_{\text{hardware}}$ is the hardware failure rate computed using traditional methods and CF is the complexity factor defined in Table 7.1. Figure 7.2 shows a sample reliability improvement failure rate function for software modeled with $\lambda_{\text{hardware}} = 1.5 \times 10^{-5}$ failures/h, CF = 100%, and $D_n = 50\%$ for all n.

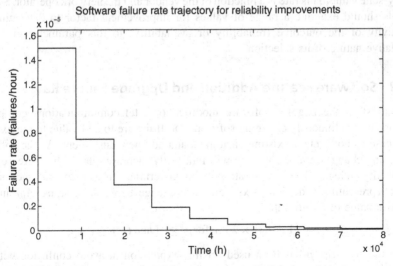

Figure 7.2. Software reliability improvement failure rate function.

The procedure outlined above is founded in industry experience showing that software reliability can play a role ranging from a relatively minor impact to the dominant factor in a system's performance. The complexity factor captures this performance impact. The hardware failure rate is used to set the initial magnitude of the software failure based on the assumption that the hardware and software elements within the device or system are treated with similar levels of rigor. Although the selection of complexity factor is qualitative, completely ignoring software failures as part of a telecommunications system model is clearly a much worse decision.

As software and firmware running on telecommunications hardware ages, bugs and faults are discovered and resolved. The interval between bug fix applications T_n depends on how responsive the equipment vendor is to repair requests and the scheduling and notification requirements of the network operator. The interval T_n can be written as the difference between two interval endpoints:

$$T_n = t_{n+1} - t_n \qquad (7.5)$$

The intervals T_n need not necessarily be integer multiples of one another. This period could be represented by random intervals representing ad hoc bug fixes, regularly spaced maintenance, or a combination of both. Practical intervals can range from days (in the case of emergency, service affecting fault repairs) to months for low-priority repairs. The improvement obtained by applying each bug fix is dependent on frequency of updates and the significance of the bugs discovered early on in the system life. Systems implementing frequent (biweekly or monthly) updates would expect to see less significant improvement with each application than a system implementing bug fixes on an annual basis. Both approaches should produce the same steady-state result in which the total software failure rate due to unresolved bugs tends to zero as time goes to infinity.

The improvement factor D_n controls how quickly the function converges on this steady-state value. This rate is a function of the vendor and the network operator. System models should examine a range of values for improvement factor to determine the sensitivity of the outcome (reliability or availability) to this parameter given the qualitative nature of its selection.

7.2.2   Software Feature Addition and Upgrade Failure Rate

In contrast to the bug fix software modifications, telecommunications equipment vendors are continuously changing software and firmware to add value to the existing code base to both retain existing customers and add new customers. As the software grows and changes, the failure rate will inherently increase due to the increase code complexity and size. The failure rate function governing this growth is a function of a factor representing code complexity and size change between versions (A_p) and the previous value of failure rate.

$$\lambda_p = \left(1 + A_p\right) \times \lambda_{p-1} \text{ for } p \geq 1 \text{ and } t_{p-1} \leq t < t_p \qquad (7.6)$$

where the subscript p has been used for this expression to avoid confusion with the reliability improvement failure rate expression and $\lambda_0 = \lambda_{FA\ Initial}$. The initial failure rate

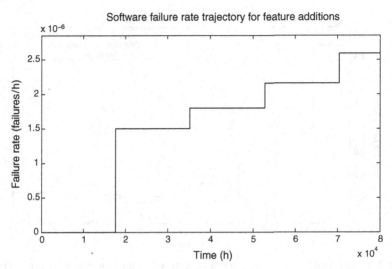

Figure 7.3. Software feature addition and upgrade failure rate function.

in this expression sets the magnitude of the software degradation due to code changes and complexity increases. Selection of this value is qualitative and requires careful consideration. Figure 7.3 shows a sample plot of $\lambda_{FA}(t)$ with $A_p = 0.2$ and $\lambda_{FA\ Initial}$ equal to 20% of the second interval reliability improvement failure rate.

Knowledge can be obtained through vendor relationships regarding the size and complexity of code releases. This knowledge can be used to set the value A_p in a manner similar to the approach discussed for the reliability improvement failure rate function. Many vendors have annual or biannual code release schedules. These release (and subsequent adoption) schedules are used to define the interval between each failure rate change in the function.

7.2.3 Total Software Failure Rate

The total software failure rate of the failure rate model is the sum of both the reliability improvement function and the feature addition and upgrade function. The resultant failure rate trajectory has an initially negative slope that becomes increasingly shallow and approaches a steady-state value as time becomes large. This general shape is consistent both qualitatively and quantitatively with the expected behavior of tele-communications-system-specific software and firmware. If we sum the two failure rate functions, we can obtain an aggregate failure rate function that includes both the ongoing reliability improvements experienced in a telecommunications system and the ongoing feature and version changes that are also an integral part of operational systems (equation 7.1 is repeated below).

$$\lambda(t) = \lambda_{RI}(t) + \lambda_{FA}(t)$$

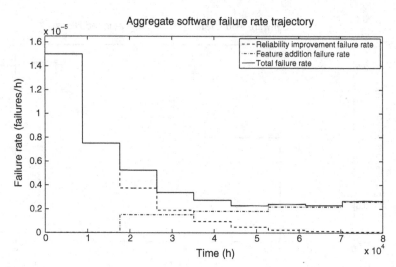

Figure 7.4. Aggregate software failure rate trajectory for reliability improvement and feature addition.

Figure 7.4 shows the individual failure rate functions as well as the total failure rate function.

Note that even though we would expect the total failure rate to tend to zero as time becomes large, the introduction of ever-changing feature sets and code size growth lead the system to settle on a steady-state, nonzero software failure rate. This result is qualitatively consistent with real-world performance of telecommunications software and firmware.

7.3 RELIABILITY AND AVAILABILITY OF SYSTEMS WITH SOFTWARE COMPONENTS

The software failure rate model can be incorporated into a complete system availability model by performing Monte Carlo simulation. Calculation of the reliability and availability metrics is performed using the same methods previously discussed. The main complication in this analysis is the time dependence of the software failure rate. Consider a system made up of two basic components: hardware and software. Figure 7.5 shows a block diagram describing the system.

The hardware elements of the component represent an aggregate failure rate for a black box of hardware items that are computed using traditional analysis methods. The software is modeled using the approach presented in this section.

7.3.1 Component Reliability

The reliability of a component, subsystem, or system using software or firmware is calculated by taking the product (serial combination) of the hardware reliability

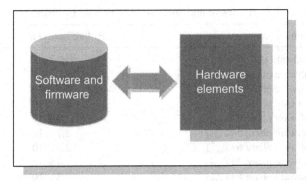

Figure 7.5. Component block diagram consisting of hardware and software.

function with the software reliability function. For the component described in Figure 7.5, the reliability can be calculated by applying

$$R(t) = R_{\text{hw}}(t) \times R_{\text{sw}}(t) \tag{7.8}$$

where $R_{\text{hw}}(t)$ and $R_{\text{sw}}(t)$ are the hardware and software reliabilities, respectively. Adopting the constant failure rate (and thus exponentially distributed TTF) assumption for both systems (as has been done previously), we can calculate the hardware and software reliability functions to be

$$R_{\text{hw}}(t) = e^{-\lambda_{\text{hw}}*t} \tag{7.9}$$

$$R_{\text{sw}}(t) = \begin{cases} e^{-\lambda_{\text{sw1}}*t} & 0 \leq t \leq t_1 \\ e^{-\lambda_{\text{sw2}}*t} & t_1 < t \leq t_2 \\ \quad\vdots \\ e^{-\lambda_{\text{swn}}*t} & t_{n-1} < t \leq t_n \end{cases} \tag{7.10}$$

Thus, the resultant reliability function is a piecewise function given by

$$R(t) = \begin{cases} e^{-(\lambda_{\text{sw1}}+\lambda_{\text{hw}})*t} & 0 \leq t \leq t_1 \\ e^{-(\lambda_{\text{sw2}}+\lambda_{\text{hw}})*t} & t_1 < t \leq t_2 \\ \quad\vdots \\ e^{-(\lambda_{\text{swn}}+\lambda_{\text{hw}})*t} & t_{n-1} < t \leq t_n \end{cases} \tag{7.11}$$

Consider a component with a stated hardware failure rate of 15×10^{-6} failures/h (67,000 h MTBF). Using the software model presented in the previous section, we have the data summarized in Table 7.2.

Figures 7.6 and 7.7 show graphs of the reliability functions for the hardware, software, and total system reliability.

Note that the reliability function for the software observes positive spikes in the reliability of the software component based on the improvements made to the failure rate at discrete intervals.

Table 7.2. Hardware and Software Failure Rate by Phase

Component	Failure Rate (Failures/h)
Hardware (λ_{hw})	15×10^{-6}
Software phase 1 (λ_{sw1})	15×10^{-6}
Software phase 2 (λ_{sw2})	7.5×10^{-6}
Software phase 3 (λ_{sw3})	5.3×10^{-6}
Software phase 4 (λ_{sw4})	3.4×10^{-6}
Software phase 5 (λ_{sw5})	2.7×10^{-6}
Software phase 6 (λ_{sw6})	2.3×10^{-6}
Software phase 7 (λ_{sw7})	2.4×10^{-6}
Software phase 8 (λ_{sw8})	2.3×10^{-6}
Software phase 9 (λ_{sw9})	2.7×10^{-6}

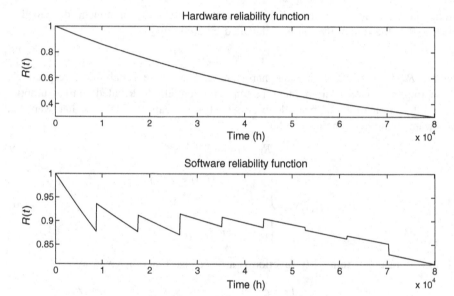

Figure 7.6. Discrete hardware and software component reliability functions.

Multiplying the hardware reliability function with the software function results in a smoothing of the software function's behavior. As expected, the reliability function becomes dominated by hardware failures as time becomes large and the impact of software failures becomes smaller with respect to the total component failure rate.

7.3.2 Component Availability

Availability metrics calculated for systems including software and hardware require the use of Monte Carlo methods. The time dependence of the software failure rate may be analyzed for simple systems involving a small number of software changes using

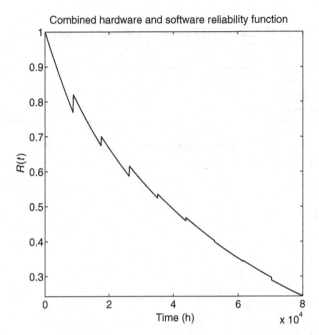

Figure 7.7. Total component reliability function for hardware and software.

closed-form methods (such as the Markov chain approach). The limitation of this approach is that adoption of an exponential distribution for TTR is required. As discussed previously, for all but the mean value case, the exponential TTR model does not produce realistic or reliable results.

The software TTR is defined in the same manner as the hardware TTR. The distribution for modeling TTR of a software component should be selected to capture a number of software failure resolution types. In many cases, software repair is most frequently (and efficiently) executed by performing a software "restart." This restart places the software in a known base (or home) state. Restarting software will frequently resolve an error or a fault condition temporarily or permanently. In some cases, the software restart will not resolve the issue. In those cases, repairing the system may take significantly longer time and may require vendor technical assistance. For this reason, a Weibull distribution is well suited to modeling software TTR. The flexibility of the Weibull distribution allows analysts to model a wide range of different conditions. Figure 7.8 shows example TTR distributions for software and hardware components. A software TTR model is shown where the majority of software repairs are executed quickly while a finite number of repairs take a significantly longer time to resolve.

The Weibull distribution shown in Figure 7.8 has a shape parameter equal to 9.0 h and a scale parameter equal to 1.8 h. The repair distribution for the hardware component is modeled using a normal distributed random variable with a mean of 8.0 h and a standard deviation of 1.5 h.

Table 7.3. Software, Hardware, and Combined Availability Performance

System Element	Availability (%)
Software element only	99.988
Hardware element only	99.991
Combined software/hardware	99.979

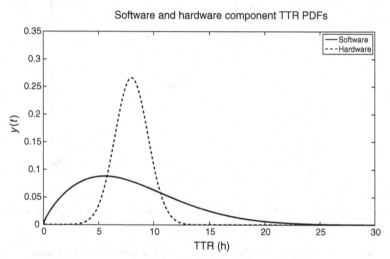

Figure 7.8. Sample software TTR distribution.

Using the TTR models presented above, the component availability can be modeled by using the Monte Carlo simulation methods presented in Chapter 2. The component is simulated for a 10-year period and 5000 system life samples are computed. The mean availability performance for the software, hardware, and combined analyses are tabulated in Table 7.3.

Figure 7.9 shows histograms of the achieved availability for the software and hardware components individually (note that the y-axis has been truncated to show detail within the nonunity region of the availability distribution).

It should be clear that the software component of the system has a distinctly different availability signature than the hardware component. Several "humps" occur in the distribution of achieved values representing the range of achieved results based on the time of failure and current failure rate of the software component. The hardware distribution follows the more traditional distribution observed in the previous analyses. The combined availability of the software and hardware is shown in Figure 7.10.

By computing the combined availability distribution of the software and hardware components, we can observe the influence of the software element on the total component's availability. In a manner similar to the reliability result, we see that the varying effects of the software elements are smoothed somewhat by the hardware component's behavior. The results of the analysis shown above are intended as an

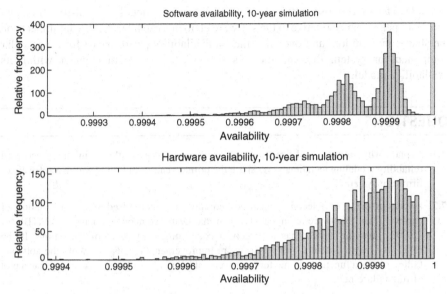

Figure 7.9. Software and hardware component availability distributions.

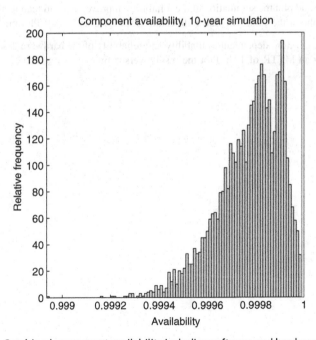

Figure 7.10. Combined component availability including software and hardware components.

example. The analyst's selection of software failure rate model, TTR distribution, and hardware performance relative to the software elements can have significant impacts on the shape, distribution, and overall achieved availability performance for a particular component or system that incorporates software as a system element within its availability model.

QUESTIONS

7.1. Explain why software and firmware cannot be modeled using the traditionally accepted reliability modeling techniques. What are the primary causes of failure within a software or firmware system?

7.2. A certain piece of telecommunications equipment is comprised of a combination of hardware and software components. The hardware components have an MTBF of 95,000 h and the system has a perceived code complexity level of "medium." If the reliability improvement factor for software improvements is 65%, compute the software failure rate as a function of time for improvement intervals of 14 months and generate a plot of this failure rate.

7.3. The same piece of equipment described in Q7.2 receives general release feature updates on an 18-month schedule. These feature additions generally increase code size by 25%. Assume that the first feature addition produces a failure rate increase equal to 25% of the current software failure rate. Plot the feature addition failure rate as a function of time.

7.4. Compute and plot the summation of the reliability improvement and feature addition failure rates to obtain the total software failure rate from Q7.2 and Q7.3. Plot the results.

7.5. Calculate the time-dependent availability and reliability of the hardware described in Q7.2 assuming an MTTR of 12 h. Plot the results versus time.

REFERENCES

Bain LJ, Englehardt M. *Introduction to Probability and Mathematical Statistics.* Boston, MA: PWS-Kent Publishing Company; 1992.

Bauer E, Zhang X, Kimber DA. *Practical System Reliability.* New York: Wiley; 2009.

Bauer E. *Design for Reliability.* Hoboken, NJ: Wiley; 2010.

DeCoster D. Diesel Start Reliability and Availability. UPS Application Paper, Mission Critical West, Inc.; C September 2010.

Garg VK. *IS-95 CDMA and CDMA2000.* Upper Saddle River, NJ: Prentice-Hall; 2000.

Hale PS Jr., Arno RG. Survey of reliability and availability information for power distribution, power generation, and HVAC components for commercial, industrial, and utility installations, *IEEE Trans. Industry Applicat.*, **37**(1): 191–196; 2001.

Hoopingarner KR, Kirkwood BJ, Louzecky PJ. Study Group Review of Nuclear Service Diesel Generator Testing and Aging Mitigation. Prepared for the U.S. Nuclear Regulatory Commission Office of Nuclear Regulatory Research Division of Engineering under Contract DE-AC06-76RLO 1830NRC/RES FIN 2911. Pacific Northwest Laboratory Operated for the U.S. Department of Energy by Battelle Memorial Institute PNL-6287; 1988.

Telecommunications System Reliability Engineering, Theory, and Practice, Mark L. Ayers.
© 2012 by the Institute of Electrical and Electronics Engineers, Inc. Published 2012 by John Wiley & Sons, Inc.

Hoopingarner KR, Zaloudek FR. *Aging Mitigation and Improved Programs for IEEE-Std 493-1997. IEEE Recommended Practice for the Design of Reliable Industrial and Commercial Power Systems.* New York: IEEE Press; 1998.

Koizumi Y, Yotsumoto K. *Power Supply System Reliability Taking Battery Failure into Account.* Tokyo, Japan: NTT Power and Building Facilities, Inc.; 2011.

Lehpamer H. *Microwave Transmission Networks Planning, Design, and Deployment.* New York: McGraw-Hill; 2010.

Liu X, Wang W. VRLA battery system reliability and proactive maintenance, IEEE 32nd INTELEC, 2010; 1–7.

McLinn J. A short history of reliability, *Reliability Review: The R&M Eng. J.* March 2010.

MIL-HDBK-472. Military Handbook: Maintainability Prediction; 1984.

MIL-STD-721C. Military Standard: Definition of Terms for Reliability and Availability; 1981.

Nuclear Service Diesel Generators. Prepared for U.S. Nuclear Regulatory Commission. Pacific Northwest Laboratory Operated by Battelle Memorial Institute PNL-6397; 1989.

Pan J. Software Reliability. Carnegie Mellon University, 18-849b Dependable Embedded Systems Spring 1999. Available at http://www.ece.cmu.edu/~koopman/des_s99/sw_reliability/.

Pritchard WL, Suyderhoud HG, Nelson RA. *Satellite Communications Systems Engineering Second Edition.* Upper Saddle River, NJ: Prentice-Hall; 1993.

Pukite J, Pukite P. *Modeling for Reliability Analysis.* New York: IEEE Press; 1998.

Rasmussen N. Calculating Total Cooling Requirements for Data Centers Revision 2007-2. White Paper #25. American Power Conversion; 2007.

Rausand M, Høyland A. *System Reliability Theory Models, Statistical Methods, and Applications*, 2nd edition. Hoboken, NJ: Wiley; 2004.

Smith D. *Reliability Maintainability and Risk.* Oxford, U.K.: Butterworth-Heinemann; 2001.

Way K, Ming JZ. *Optimal Reliability Modeling: Principles and Applications.* Hoboken, NJ: Wiley; 2003.

INDEX

Telecommunications System Reliability Engineering, Theory, and Practice, Mark L. Ayers.
© 2012 by the Institute of Electrical and Electronics Engineers, Inc. Published 2012 by John Wiley & Sons, Inc.

IEEE Press Series in
Network Management

The goal of this series is to publish high quality technical reference books and textbooks on network and services management for communications and information technology professional societies, private sector and government organizations as well as research centers and universities around the world. This Series focuses on Fault, Configuration, Accounting, Performance, and Security (FCAPS) management in areas including, but not limited to, telecommunications network and services, technologies and implementations, IP networks and services, and wireless networks and services.

Series Editors:
Thomas Plevyak
Veli Sahin

Printed in the United States
By Bookmasters